CHICAGO PUBLIC LIBRARY
HAROLD WASHINGTON LIBRARY CENTER

D0853656

ding

HILL'S EQUATION

WILHELM MAGNUS
STANLEY WINKLER

DOVER PUBLICATIONS, INC.
NEW YORK

Published in Canada by General Publishing Company, Ltd., 30 Lesmill Road, Don Mills, Toronto, Ontario.

Published in the United Kingdom by Constable and Company, Ltd., 10 Orange Street, London WC2H 7EG.

This Dover edition, first published in 1979, is an unabridged and corrected republication of the work originally published by John Wiley & Sons, New York, in 1966. The authors have contributed a new preface and prepared a list of additional references for this Dover edition.

International Standard Book Number: 0-486-63738-7
Library of Congress Catalog Card Number: 78-74114

Manufactured in the United States of America
Dover Publications, Inc.
180 Varick Street
New York, N.Y. 10014

PREFACE TO THE DOVER EDITION

Since the publication of the first edition of *Hill's Equation* in 1966, a large number of important theorems have been proved which connect the periodic function $Q(x)$ in Hill's equation with the spectrum of the equation, i.e., with the sequence

$$\lambda_0', \lambda_1', \lambda_2, \lambda_1, \lambda_2, \ldots$$

of its characteristic values. To begin with, Borg's theorem (page 47) can now be proved briefly with function theoretic methods [Hochstadt (1977)] and can be supplemented by a characterization of those $Q(x)$ for which all even finite intervals of instability disappear [Hochstadt (1978)]. The result of Hochstadt (1963b) that the intervals of instability decrease rapidly if $Q(x)$ is in C_1^{∞}, has been sharpened and inverted by McKean and Trubowitz (1976) and by Trubowitz (1977). The latter paper proves that $Q(x)$ is analytic if and only if the widths of the intervals of instability are exponentially decreasing. Other interesting results connecting $Q(x)$ and the spectrum which, however, cannot be characterized in a few words, are contained in Hochstadt (1966, 1968, 1975, 1976a and b) and Trubowitz (1977).

The last two general theorems quoted (but not proved) in the present book (p. 113) are due to Hochstadt (1965) and refer to the case where only a finite number N of finite instability intervals exists. For $N = 1$, Hochstadt had shown that Q satisfies a second order nonlinear differential equation which, incidentally, shows that Q must be an elliptic function. For $N > 1$, Flaschka (1975) and, independently, Goldberg (1975, 1976) showed that Q must satisfy a nonlinear differential equation of order $2N$ involving $N + 1$ arbitrary parameters which can, for any fixed N, be written out explicitly with the help of simple recursion formulas. The importance of this result is greatly enhanced by the striking discovery of Lax (1975) according to which the same functions Q appear in the construction of certain solutions of the Korteweg-deVries equation

$$u_t + uu_x + u_{xxx} = 0$$

which are periodic in x and almost periodic in t. More precisely, the results of Flaschka and Goldberg establish the fact that exactly all those Hill's equations which have finitely many intervals of instability can be obtained from functions Q which appear in the construction of periodic solutions of the Korteweg-deVries equation.

We cannot describe here in detail the connection between Hill's equation and the Korteweg-deVries equation. Even less can we indicate the scope of the very comprehensive new theory which appears in papers by McKean and van Moerbecke (1975) and McKean and Trubowitz (1976). We merely hope that the papers quoted under the heading "Additional References" included in this new edition will provide an adequate basis for information about the present state of knowledge. The only alteration we have made in the text is the correction of a few errata which have come to our attention.

Our sincere thanks are due to Dover Publications for making this new edition possible.

Polytechnic Institute of New York — *Wilhelm Magnus*
I.B.M. Corporation, Armonk, New York — *Stanley Winkler*

PREFACE TO THE FIRST EDITION

The term "Hill's equation" is a convenient abbreviation defining the class of homogeneous, linear, second-order differential equations with real, periodic coefficients. Although such differential equations have been investigated before the publication of the memoir on the motion of the lunar perigee by G. W. Hill in 1877, it seems to be justified to give his name to this whole class of differential equations in view of the important and lasting contributions made by Hill to their theory.

There exist hundreds of applications of Hill's equation to problems in engineering and physics, including problems in mechanics, astronomy, the theory of electric circuits, of the electric conductivity of metals, and of the cyclotron. The fundamental importance of Hill's equation for stability problems was established by Liapounoff in 1907. It may be appropriate to remark here that the theory of Hill's equation reveals the occurrence of a surprising phenomenon which can be described in rather simple terms. If a force varying periodically with time acts on a mass in such a manner that the force tends to move the mass back into a position of equilibrium in proportion to the dislocation of the mass, one might expect the mass to be confined to a neighborhood of the position of equilibrium. In particular, once the force is strong enough to achieve this effect, one would expect a stronger force to be even more efficient for this purpose. Such, however, need not be the case. In fact, an increase of the restraining force may cause the mass to oscillate with wider and wider amplitudes. The theory of the intervals of instability of Hill's equation provides the precise description of this phenomenon.

In order to describe the purpose, scope, and character of this tract, it may be said that we intend to provide orientation but had to abstain from giving full information. The elementary facts of the theory are proved in full except for those which belong to the general theory of linear differential equations. These are merely stated, and a reference is given. The mathematical tools used in the proofs are of a rather limited nature. We apply the result from the theory of entire analytic functions which states that a function of order of growth 1/2 has infinitely many zeros, and we use the Riemann-Lebesgue Theorem, but nothing more sophisticated than these fairly elementary facts. Results which can be arrived at only through very lengthy proofs have been stated but the proofs have been replaced by references. Also, we did not try to duplicate the monographs by Starzinskiĭ and Krein which deal with closely related subjects. References to these surveys are given wherever this seemed to be appropriate.

Our monograph consists of two parts, the first of which (Chapters I and II) deals with the basic theory whereas the second part (Chapters III–VIII) contains details, refinements, and special cases. As for the applications, we hope to present eventually as a separate book a survey of several hundred papers in which Hill's equation is used as a mathematical tool. To include the applications here would have more than doubled the size of the book. The same would have happened if we had tried to treat Hill's equation merely as a special case of a system of linear homogeneous differential equations with periodic coefficients. A glance at the literature quoted in the monograph by Cesari will confirm this statement.

We have tried to give credit wherever it belongs, but it is almost an *a priori* certainty that we did not succeed here. We regret any omissions and misplacements of references which may have occurred.

Our monograph is based on several reports published by the Division of Electromagnetic Research of the Institute of Mathematical Sciences of New York University. We acknowledge gratefully the sponsorship of the Mathematics Division of the Air Force Office of Scientific Research for these reports. However, we have tried to unify these earlier presentations and to supplement them with more recent results, in particular where these provided a simplification of existing proofs like some of the papers by Hochstadt.

Apart from the index, we have included a brief list of standard notations used in the book as an aid for the reader. Also, there is a list of lemmas and theorems with a reference to the page where they are stated.

New York City
Washington, D.C.
Spring, 1966

Wilhelm Magnus
Stanley Winkler

CONTENTS

PART I. GENERAL THEORY

PART II. DETAILS

PART I

GENERAL THEORY

I

Basic Concepts

1.1. *Preliminary remarks*

Any homogeneous linear differential equation of second order with real periodic coefficients can be reduced to an equation of Hill's type. A specific question which arises in the theory of Hill's equation is the problem of the existence of periodic solutions. This problem has many features in common with the ordinary Sturm-Liouville problems, and in certain cases it can, in fact, be reduced to ordinary boundary value problems of the Sturm-Liouville type (see Section 1.3). However, in general, such a reduction is not possible, and imposing the periodicity requirements on a solution of the differential equation leads to phenomena different from those resulting from the imposition of a homogeneous boundary condition of the Sturm-Liouville type. Thus, the differential equation can have two linearly independent periodic solutions but it cannot have two linearly independent solutions satisfying the same homogeneous boundary conditions. Furthermore, the value of the period of the solution (which is a multiple of the period p of the coefficients) plays an important role in the discussion of periodic solutions. In a certain sense, only the solutions of period p and $2p$ are of interest. We shall now proceed with a detailed presentation of some basic theorems and their proofs.

As references to the general theory of Sturm-Liouville (self-adjoint) boundary-value problems we mention Courant and Hilbert (1953) and Coddington and Levinson (1955).

1.2. *Floquet's theorem*

Let $Q(x)$ be a (real or complex valued) function of the real variable x defined for all values of x. We assume that $Q(x)$ is piecewise continuous in every finite interval and that it is periodic with minimal period π. By this we mean that for all x

$$Q(x + \pi) = Q(x), \tag{1.1}$$

and that if p is a number with $0 < p < \pi$, then there exists at least one real interval I such that $Q(x + p) \neq Q(x)$ for $x \in I$.

If $Q(x)$ has the properties stated above, then the differential equation

$$y'' + Q(x)y = 0 \tag{1.2}$$

has two continuously differentiable solutions $y_1(x)$ and $y_2(x)$ which are uniquely determined by the conditions:

$$y_1(0) = 1, \quad y_1'(0) = 0, \quad y_2(0) = 0, \quad y_2'(0) = 1.$$

These solutions are referred to as *normalized solutions* of (1.2).

Before stating Floquet's theorem we must define the notions of *characteristic equation* and *characteristic exponent* associated with (1.2). Thus, the characteristic equation is the equation

$$\rho^2 - [y_1(\pi) + y_2'(\pi)]\rho + 1 = 0 \tag{1.3}$$

and the characteristic exponent α is a number which satisfies the equations

$$\exp i\alpha\pi = \rho_1, \quad \exp(-i\alpha\pi) = \rho_2, \tag{1.4}$$

where ρ_1 and ρ_2 are the roots of the characteristic equation (1.3).

It is clear that α is defined up to an integral multiple of 2. Also $2 \cos \alpha\pi = y_1(\pi) + y_2'(\pi)$. Finally $\rho_1\rho_2 = 1$. We can now state the theorem.

Floquet's theorem. 1. *If the roots ρ_1 and ρ_2 of the characteristic equation* (1.3) *are different from each other, then Hill's equation* (1.2) *has two linearly independent solutions*

$$f_1(x) = e^{i\alpha x}p_1(x), \quad f_2(x) = e^{-i\alpha x}p_2(x),$$

where $p_1(x)$ and $p_2(x)$ are periodic with period π.

2. *If $\rho_1 = \rho_2$, then equation* (1.2) *has a nontrivial solution which is periodic with period π (when $\rho_1 = \rho_2 = 1$) or 2π (when $\rho_1 = \rho_2 = -1$). Let*

$p(x)$ denote such a periodic solution and let $y(x)$ be another solution linearly independent of $p(x)$. Then

$$y(x + \pi) = \rho_1 y(x) + \theta p(x), \qquad \theta \text{ constant},$$

and $\theta = 0$ is equivalent to

$$y_1(\pi) + y_2'(\pi) = \pm 2, \quad y_2(\pi) = 0, \quad y_1'(\pi) = 0.$$

Before starting with the proof of Floquet's theorem it may be appropriate to discuss its significance. Thus, let $\rho_1 \neq \rho_2$. If α is real, then there exists an upper bound M for the absolute value $|y(x)|$ of *every* solution of (1.2) and M depends only on the initial conditions for y and not on x. If α is not real, then there exists a nontrivial unbounded solution $y(x)$ of (1.2). If $\rho_1 = \rho_2$, then for all solutions of (1.2) to be bounded it is necessary and sufficient that

$$y_1(\pi) + y_2'(\pi) = \pm 2, \quad y_2(\pi) = 0, \quad y_1'(\pi) = 0.$$

Whenever all solutions of (1.2) are bounded we say that they are *stable*; otherwise we say that they are *unstable*.

The solutions of period π and 2π play an exceptional role as is seen from the following:

Corollary to Floquet's theorem. *If (1.2) has a periodic nontrivial solution with period $n\pi$, $n > 2$, but no solution with period π or 2π, then all solutions are periodic with period $n\pi$.*

Indeed, our assumption implies that $\rho_1 \neq \rho_2$ so that every solution y of (1.2) is of the form

$$y = \mu f_1(x) + \nu f_2(x).$$

If one such solution is periodic with period $n\pi$ then $y(x + n\pi) = \mu c f_1 + \nu \bar{c} f_2 = y(x)$ where $c = \exp(i\alpha n\pi)$, $\bar{c} = \exp(-i\alpha n\pi)$. Since f_1 and f_2 are linearly independent, $c = \bar{c} = 1$. Therefore, $n\alpha$ is an even integer, and both f_1 and f_2 are periodic with period $n\pi$.

Proof of Floquet's theorem. If $y(x)$ is a solution of (1.2), then, obviously, $y(x + \pi)$ is also a solution of (1.2). In particular, $y_1(x + \pi)$ and $y_2(x + \pi)$ are solutions of (1.2). Since $y_1(x)$ and $y_2(x)$ form a basis for the set of all solutions of (1.2), it must be possible to express $y_1(x + \pi)$

and $y_2(x + \pi)$ as linear combinations of $y_1(x)$ and $y_2(x)$. We find easily that

$$\begin{aligned} y_1(x + \pi) &= y_1(\pi)y_1(x) + y_1'(\pi)y_2(x) \\ y_2(x + \pi) &= y_2(\pi)y_1(x) + y_2'(\pi)y_2(x). \end{aligned} \tag{1.5}$$

Assume now that $y(x) \not\equiv 0$ is a solution of (1.2) such that

$$y(x + \pi) = \rho y(x) \tag{1.6}$$

for some constant ρ. If $y(x) = c_1y_1(x) + c_2y_2(x)$, then it follows from (1.6) that c_1 and c_2 must satisfy the system of linear equations

$$\begin{aligned} (y_1(\pi) - \rho)c_1 + y_2(\pi)c_2 &= 0 \\ y_1'(\pi)c_1 + (y_2'(\pi) - \rho)c_2 &= 0. \end{aligned} \tag{1.7}$$

Conversely, if (1.7) is satisfied, $y(x)$ satisfies (1.6). Now, the necessary and sufficient condition for (1.7) to have a solution c_1, c_2 such that c_1 and c_2 do not both vanish is

$$\begin{vmatrix} y_1(\pi) - \rho, & y_2(\pi) \\ y_1'(\pi), & y_2'(\pi) - \rho \end{vmatrix} = 0. \tag{1.8}$$

Since, for all x, the Wronskian

$$y_1(x)y_2'(x) - y_2(x)y_1'(x) = 1,$$

equation (1.8) is identical with the characteristic equation (1.3). Thus, if $\rho = \rho_1$ is a root of (1.8), we can find c_1 and c_2 such that $y = c_1y_1 + c_2y_2 \not\equiv 0$ and such that y satisfies (1.6). Obviously, if (1.6) is satisfied, we may write

$$y = y(x) = \exp(i\alpha x)p_1(x) = f_1(x)$$

where $\exp(i\alpha\pi) = \rho_1$ and where $p_1(x)$ is a periodic function of x with period π. Suppose now that (1.8) has a second solution $\rho = \rho_2 \neq \rho_1$. We may use ρ_2 for the construction of a solution $y = f_2(x) \not\equiv 0$ of (1.3) such that $f_2(x + \pi) = \rho_2 f_2(x)$. We observe that f_1 and f_2 are linearly independent. Otherwise, we could find constants λ_1 and λ_2 (both not equal to zero) such that

$$\lambda_1 f_1(x) + \lambda_2 f_2(x) \equiv 0.$$

But then we would also have

$$\lambda_1 f_1(x + \pi) + \lambda_2 f_2(x + \pi) = \lambda_1\rho_1 f_1(x) + \lambda_2\rho_2 f_2(x) \equiv 0.$$

Since both $\lambda_1 f_1$ and $\lambda_2 f_2$ do not vanish identically, the last three equations are compatible only if $\rho_1 = \rho_2$, which we have excluded. This proves Floquet's theorem in the case where $\rho_1 \neq \rho_2$.

Since $\rho_1\rho_2 = 1$, either $|\rho_1| = |\rho_2| = 1$ or at least one of the numbers $|\rho_1|$ or $|\rho_2|$ exceeds 1. In the first case, we have stability, in the second case, instability of the solutions of (1.2), provided that $\rho_1 \neq \rho_2$.

If $\rho_1 = \rho_2$, we still can construct one solution $y_1^*(x)$ of (1.2) such that

$$y_1^*(x + \pi) = \rho_1 y_1^*(x).$$

Since $\rho_1 = \rho_2$ and $\rho_1\rho_2 = 1$ imply that $\rho_1 = \pm 1$, y_1^* is obviously periodic with period π or 2π. In order to find the properties of a solution $y_2^*(x)$ which is linearly independent of y_1^*, assume first that $y_2(\pi) \neq 0$. Then we may choose [cf. (1.7) and (1.3)]

$$y_1^*(x) = y_2(\pi)y_1(x) + [\rho_1 - y_1(\pi)]y_2(x)$$
$$y_2^*(x) = y_2(x)$$

and we find from $2\rho_1 = y_1(\pi) + y_2'(\pi)$ that

$$y_2^*(x + \pi) = \rho_1 y_2^*(x) + y_1^*(x).$$

Similarly, if $y_2(\pi) = 0$, we may choose

$$y_1^*(x) = y_2(x), \quad y_2^*(x) = y_1(x).$$

Since $y_1(\pi)y_2'(\pi) - y_1'(\pi)y_2(\pi) = 1$ it follows from $y_2(\pi) = 0$ and $y_1(\pi) + y_2'(\pi) = 2\rho_1$ than $y_1(\pi) = y_2'(\pi) = \rho_1$ and therefore we have from (1.5) that

$$y_1^*(x + \pi) = \rho_1 y_1^*(x)$$
$$y_2^*(x + \pi) = \rho_1 y_2^*(x) + y_1'(\pi)y_1^*(x).$$

This proves Floquet's theorem in all details.

As a rather obvious consequence of Floquet's theorem we mention the following

Stability test. *The solutions of* (1.2) *are stable if and only if* $y_1(\pi) + y_2'(\pi)$ *is real and*

$$|y_1(\pi) + y_2'(\pi)| < 2$$

or

$$y_1(\pi) + y_2'(\pi) = \pm 2$$

and

$$y_2(\pi) = y_1'(\pi) = 0.$$

Proof. If $\rho_1 \neq \rho_2$, then stability is equivalent to $\alpha \neq 0$, α real, which, in turn, is equivalent to $y_1(\pi) + y_2'(\pi)$ being real and in absolute value < 2. If $\rho_1 = \rho_2$, then stability is equivalent to $y_1(\pi) + y_2'(\pi) = \pm 2$ and $y_2(\pi) = y'_1(\pi) = 0$.

1.3. The symmetric case $Q(x) = Q(-x)$

If in (1.2) the function $Q(x)$ is even, i.e., if

$$Q(x) = Q(-x) \tag{1.9}$$

it is possible to establish relations between the values of y_1, y_2, y_1', and y_2' at $x = \pi/2$ and at $x = \pi$ and these relations allow a more detailed classification of the solutions of period π and 2π. We summarize our results by stating

Theorem 1.1. *Let $y_1(x)$ and $y_2(x)$ be the normalized solutions of* (1.2) *and assume that $Q(x)$ satisfies* (1.9). *Then the following relations hold:*

$$y_1(\pi) = 2y_1(\pi/2)y_2'(\pi/2) - 1 = 1 + 2y_1'(\pi/2)y_2(\pi/2) \tag{1.10}$$

$$y_2(\pi) = 2y_2(\pi/2)y_2'(\pi/2) \tag{1.11}$$

$$y_1'(\pi) = 2y_1(\pi/2)y_1'(\pi/2) \tag{1.12}$$

$$y_2'(\pi) = y_1(\pi). \tag{1.13}$$

In all cases, $y_1(x) = y_1(-x)$, i.e., $y_1(x)$ will be an even function of x. Similarly, $y_2(x) = -y_2(-x)$ will be an odd function. Whenever a nontrivial solution of period π or 2π exists, there also exists such a solution which is either odd or even. Therefore, these periodic solutions are necessarily multiples of one of the normalized solutions $y_1(x)$ or $y_2(x)$ unless all solutions are periodic (with period π or 2π).

Theorem 1.2. *If the conditions of Theorem* 1.1 *are satisfied, then there exists a nontrivial periodic solution of* (1.2) *which is*

(1) *even and of period π if and only if $y_1'(\pi/2) = 0$*

(2) *odd and of period π if and only if $y_2(\pi/2) = 0$*

(3) *even and of period 2π if and only if $y_1(\pi/2) = 0$*

(4) *odd and of period 2π if and only if $y_2'(\pi/2) = 0$.*

Periodic solutions of period π or 2π are necessarily multiples of the normalized solutions $y_1(x)$ and $y_2(x)$.

Proof of Theorem 1.1. If $Q(x)$ is even and if $y(x)$ is a solution of (1.2), then $y(-x)$ is also a solution. Since the initial conditions for $y_1(-x)$ and $y_1(x)$ coincide and, similarly, those for $y_2(x)$ and $-y_2(-x)$ are identical, it follows that $y_1(x)$ is even and $y_2(x)$ is odd. Therefore, we find from (1.5) for $x = -\pi/2$:

$$y_1(\pi/2) = y_1(\pi)y_1(\pi/2) - y_1'(\pi)y_2(\pi/2) \tag{1.14}$$

$$y_2(\pi/2) = y_2(\pi)y_1(\pi/2) - y_2'(\pi)y_2(\pi/2). \tag{1.15}$$

Obviously, $y_1'(x)$ is odd and $y_2'(x)$ is even. Using this fact, we find from (1.5) by differentiating both sides with respect to x and by putting $x = -\pi/2$ that

$$y_1'(\pi/2) = -y_1(\pi)y_1'(\pi/2) + y_1'(\pi)y_2'(\pi/2) \tag{1.16}$$

$$y_2'(\pi/2) = -y_2(\pi)y_1'(\pi/2) + y_2'(\pi)y_2'(\pi/2). \tag{1.17}$$

We may treat (1.14) to (1.17) as a system of linear equations for $y_1(\pi)$, $y_2(\pi)$, $y_1'(\pi)$, and $y_2'(\pi)$. Solving the equations for these quantities, we arrive at Theorem 1.1 if we observe that

$$y_1(\pi/2)y_2'(\pi/2) - y_1'(\pi/2)y_2(\pi/2) = 1.$$

If $y(x)$ is of period π or 2π, the functions

$$u(x) = y(x) + y(-x), \quad v(x) = y(x) - y(-x)$$

have the same property. Since u is even and v is odd, and since u and v cannot both be trivial unless $y \equiv 0$, the last statement of Theorem 1.1 follows.

Proof of Theorem 1.2. We shall now use the fact that an even solution of (1.2) is a multiple of $y_1(x)$ and that an odd solution must be a multiple of $y_2(x)$.

To prove (1) let us assume that $y(x)$ is a nontrivial, even, periodic solution of (1.2) with period π. Then $y_1(x)$ is also periodic with period π and the same is true about $y_1'(x)$. Thus $y_1'(\pi/2) = y_1'(-\pi/2)$. Since $y_1'(x)$ is an odd function it follows that $y_1'(\pi/2) = -y_1'(-\pi/2)$. Consequently, $y_1'(\pi/2) = 0$. Conversely, if $y_1'(\pi/2) = 0$, then $y_1'(-\pi/2) = 0$.

Also, $y_1(-\pi/2) = y_1(\pi/2)$. Therefore $y_1(x)$ satisfies the same conditions at $x = -\pi/2$ and at $x = \pi/2$. From this and from the periodicity of $Q(x)$ it follows that $y_1(x)$ is periodic with period π.

The proof of (2) is entirely analogous to the proof of (1).

In proving (3) we show that the values of $y_1(x)$ and $y_1'(x)$ at $x = -\pi/2$ differ in sign from the values of $y_1(x)$ and $y_1'(x)$ at $x = \pi/2$. This shows that $y_1(x + \pi) = -y_1(x)$ and therefore $y_1(x + 2\pi) = y_1(x)$.

The proof of (4) is analogous to the proof of (3).

II

Characteristic Values and Discriminant in the Real Case

2.1. *Characteristic values and intervals of stability*

In this and in the following sections we shall study Hill's equation in its standard form

$$y'' + [\lambda + Q(x)]y = 0, \tag{2.1}$$

where λ is a parameter and where $Q(x)$ is a *real* periodic function of x with period π.

Unless otherwise stated, we shall assume that $Q(x)$ is of bounded variation. In some cases, we shall even assume that $Q(x)$ is twice differentiable for all x. Usually, the proofs of the theorems indicate clearly which assumptions about $Q(x)$ one has to make.

Let y_1 and y_2 be the two linearly independent solutions of (2.1) which we defined by simple initial conditions in Section 1.2. To emphasize their dependence on λ we shall sometimes write $y_1(x, \lambda)$ and $y_2(x, \lambda)$ instead of $y_1(x)$ and $y_2(x)$. One of our main problems will be the determination of those values of λ for which the solutions of (2.1) are stable. Another one will be the problem of determining those values of λ for which equation (2.1) has a solution of period π or 2π. The following theorem due to Liapounoff (1907) and to Haupt (1914, 1918) connects the two problems:

Theorem 2.1. (*Oscillation Theorem*) *To every differential equation* (2.1), *there belong two monotonically increasing infinite sequences of real numbers*

$$\lambda_0, \quad \lambda_1, \quad \lambda_2 \quad \ldots \tag{2.2}$$

and

$$\lambda_1', \quad \lambda_2', \quad \lambda_3', \quad \lambda_4', \quad \ldots \tag{2.3}$$

such that (2.1) *has a solution of period* π *if and only if* $\lambda = \lambda_n$, $n = 0$, 1, 2, ..., *and a solution of period* 2π *if and only if* $\lambda = \lambda_n'$, $n = 1$, 2, 3, ... *The* λ_n *and* λ_n' *satisfy the inequalities*

$$(2.4) \quad \lambda_0 < \lambda_1' \leq \lambda_2' < \lambda_1 \leq \lambda_2 < \lambda_3' \leq \lambda_4' < \lambda_3 \leq \lambda_4 < \cdots$$

and the relations

$$(2.5) \qquad \lim_{n\to\infty} \lambda_n^{-1} = 0, \quad \lim_{n\to\infty} (\lambda_n')^{-1} = 0.$$

The solutions of (2.1) *are stable in the intervals*

$$(2.6) \qquad (\lambda_0, \lambda_1'), \quad (\lambda_2', \lambda_1), \quad (\lambda_2, \lambda_3'), \quad (\lambda_4', \lambda_3), \quad \ldots.$$

At the endpoints of these intervals the solutions of (2.1) *are, in general, unstable. This is always true for* $\lambda = \lambda_0$. *The solutions of* (2.1) *are stable for* $\lambda = \lambda_{2n+1}$ *or* $\lambda = \lambda_{2n+2}$ *if and only if* $\lambda_{2n+1} = \lambda_{2n+2}$, *and they are stable for* $\lambda = \lambda'_{2n+1}$ *or* $\lambda = \lambda'_{2n+2}$ *if and only if* $\lambda'_{2n+1} = \lambda'_{2n+2}$.

For complex values of λ (2.1) *has always unstable solutions.*

The λ_n *are the roots of the equation* $\Delta(\lambda) = 2$ *and the* λ_n' *are those of* $\Delta(\lambda) = -2$, *where*

$$\Delta(\lambda) = y_1(\pi, \lambda) + y_2'(\pi, \lambda).$$

In order to refer briefly to the assertions of Theorem 2.1, we shall use the following definitions:

The real numbers λ_n shall be called *characteristic values of the first kind* of (2.1) and the λ_n' shall be called *characteristic values of the second kind.* The intervals (2.6) on the real λ axis shall be called *intervals of stability*; an end-point of such an interval shall belong to it if and only if (2.1) has stable solutions for the corresponding value of λ. Similarly, we shall talk about *intervals of instability.* The intervals of both stability and of instability are ordered in a natural manner. The interval of instability $(-\infty, \lambda_0)$ will always be present. We shall call it the zeroth interval of instability and we shall call (λ_1', λ_2') the first interval of instability. Observe that, according to Theorem 2.1, neither an interval of stability nor an interval of instability can ever shrink to a point. The intervals of stability can never disappear, but two of them can combine to a single one if $\lambda_{2n+1} = \lambda_{2n+2}$ or $\lambda'_{2n+1} = \lambda'_{2n+2}$. However, the intervals of instability (with the exception of the zeroth interval) may disappear altogether. This takes place if $Q(x)$ is a constant. According to Borg (1946), this is also the only case for such an occurrence; a simple proof of this fact, due to Ungar (1961), will be given in Section 7.6.

Lamé's equation (Section 7.3) offers an example where only a prescribed, finite, number of intervals of instability remains. That Lamé's equation is the only case where only two intervals of instability remain is a recent result of Hochstadt (1965). (See Section 7.6.)

Proof of Theorem 2.1. We shall exclude first the possibility of stable solutions of (2.1) in the case of a complex value of λ. Assume that $\lambda = \mu + i\nu$, where μ and ν are real and $\nu \neq 0$. Let $y = u + iv$ be a solution of (2.1) which is of the type

$$y(x) = e^{i\alpha x}p(x) = u + iv \tag{2.7}$$

where α is a real and where $p(x)$ is periodic with period π. According to Floquet's theorem, such a solution $y(x)$ exists if we have stability for the solutions of (2.1). By splitting (2.1) into its real and imaginary parts, we find

$$\begin{aligned} u'' + [\mu + Q(x)]u &= \nu v \\ v'' + [\mu + Q(x)]v &= -\nu u. \end{aligned} \tag{2.8}$$

If we multiply the first of the equations (2.8) by v and the second by u and subtract, we see that

$$u''v - v''u = \nu(u^2 + v^2),$$

or, upon integrating, that

$$u'v - v'u = \nu \int_0^x [u^2(t) + v^2(t)]\,dt + c, \tag{2.9}$$

where c is a constant. Now we see from (2.7) that all of the functions $|u|$, $|v|$, $|u'|$, and $|v'|$ must be bounded for all values of x (since $p(x)$ is a differentiable periodic function of x). Therefore, there exists an upper bound for $|uv' - vu'|$ which is independent of x. According to (2.9), the same must be true for the absolute value of

$$I(x) = \int_0^x [u^2(t) + v^2(t)]\,dt.$$

However, $|I(x)| \to \infty$ as $x \to \infty$ since $u^2 + v^2 = |p|^2$ and therefore, for $n = 1, 2, 3, \ldots$,

$$I(n\pi) = \int_0^{n\pi} [u^2(t) + v^2(t)]\,dt = n \int_0^{\pi} |p(t)|^2\,dt.$$

Therefore if λ is not real we cannot have a solution of type (2.7).

Next we wish to show that there exists a real number λ^* such that for any $\lambda \leq \lambda^*$ the solutions of (2.1) are unstable. For this purpose, select a λ^* such that for all x

$$\lambda^* + Q(x) < 0.$$

This is certainly possible since $Q(x)$, being periodic, is a bounded function of x. We shall show that if $\lambda \leq \lambda^*$ then $y_1(x, \lambda) \to \infty$ as $x \to \infty$. To this end we shall write (2.1) in the form

$$y'' = D(x)y \tag{2.10}$$

where $D(x) = -\lambda - Q(x) > 0$ for all x. Since $y_1(0) = 1$ and $y_1''(0) > 0$ it follows from $y_1'(0) = 0$ that $y_1'(x) > 0$ for all sufficiently small positive x. Therefore, if the set S of positive zeros of $y_1'(x)$ is not empty it has a greatest lower bound $\varepsilon > 0$.

We shall show that ε does not exist and that therefore $y_1'(x) > 0$ for all $x > 0$. For this purpose we observe that the continuity of $y_1'(x)$ implies that $y_1'(\varepsilon) = 0$. Now we have from (2.10) that

$$(y_1'(\varepsilon))^2 = 2\int_0^\varepsilon D(x)y_1(x)y_1'(x)\,dx. \tag{2.11}$$

Since $y_1(0) = 1$ and $y_1'(x) \geq 0$ for $0 \leq x \leq \varepsilon$ we have $y_1(x) > 0$, $D(x) > 0$, and $y_1'(x) > 0$ for $0 < x < \varepsilon$. Therefore, the right-hand side of (2.11) is positive whereas we had assumed that $y_1'(\varepsilon) = 0$. This shows that $y_1'(x) > 0$ if $x > 0$. Therefore $y_1(x)$ is monotonically increasing for $x > 0$ which means that $y_1(x) \geq 1$ for $x > 0$. Since $y_1'(x) > 0$, we see from (2.11) that $y_1'(x)$ increases monotonically with x, and since $\delta = y_1'(x_0) > 0$ for a certain $x_0 > 0$, we see that

$$y_1(x) \geq 1 + (x - x_0)\delta \qquad (x \geq x_0).$$

Therefore $y_1(x) \to \infty$ as $x \to \infty$. By a similar argument, we can show that $y_2'(x) > 1$ for $x > 0$. We have thus proved incidentally

Lemma 2.1. If $\lambda \leq \lambda^*$, then, for $x > 0$,

$$y_1(x, \lambda) + y_2'(x, \lambda) > 2.$$

Since this result implies that $\rho_1 \neq \rho_2$, we conclude that α in $y_1(x) = Ae^{i\alpha x}p_1(x) + Be^{-i\alpha x}p_2(x)$ cannot be real. We thus find that if λ is complex or if $\lambda \leq \lambda^*$, then there exists no solution $y(x)$ of (2.1) which is of type (2.7) with real α.

We shall now examine closely the properties of the functions $\Delta(\lambda) - 2$ and $\Delta(\lambda) + 2$, where

$$\Delta(\lambda) = y_1(\pi, \lambda) + y_2'(\pi, \lambda). \tag{2.12}$$

We first note that $\Delta(\lambda) = 2$ is equivalent to $\rho_1 = \rho_2 = 1$ and that $\Delta(\lambda) = -2$ is equivalent to $\rho_1 = \rho_2 = -1$. Hence, if $\Delta(\lambda) = \pm 2$, (2.1) will have a solution of type (2.7) with real α (cf. Floquet's theorem). Since λ complex or $\lambda \leq \lambda^*$ implies that (2.1) has no such solution it follows that

Lemma 2.2. All roots of the equations $\Delta(\lambda) - 2 = 0$ and $\Delta(\lambda) + 2 = 0$ are real and $> \lambda^*$.

In Section 2.2 we shall prove that both $\Delta(\lambda) - 2$ and $\Delta(\lambda) + 2$ are entire analytic functions of λ which have infinitely many zeros. According to Lemma 2.2, all of these zeros are real and greater than λ^*. This establishes the existence of the two sequences (2.2) and (2.3) of Theorem 2.1. In fact, we have immediately from Floquet's theorem that the following assertion is true:

Lemma 2.3. Equation (2.1) has a periodic solution of period π if and only if $\Delta(\lambda) = 2$ and a periodic solution of period 2π if and only if $\Delta(\lambda) = -2$.

To prove Lemma 2.3, we merely have to use once more the fact that the condition $\rho_1 = \rho_2 = 1$ of Floquet's theorem is identical with the condition $\Delta(\lambda) = 2$, and that $\rho_1 = \rho_2 = -1$ is equivalent to $\Delta(\lambda) = -2$.

Since $\Delta(\lambda)$ is an entire analytic function, the limit relations (2.5) are obviously true. We now turn to a proof of the inequalities (2.4). For this purpose, we need

Lemma 2.4. Let μ be a root of the equation $\Delta(\lambda) - 2 = 0$ such that the derivative $\Delta'(\lambda)$ of $\Delta(\lambda)$ with respect to λ is negative or zero for $\lambda = \mu$. Then $\Delta'(\lambda) < 0$ in any open interval $\mu < \lambda < \mu_1^*$ in which $\Delta(\lambda) > -2$. Similarly, let μ' be a root of $\Delta(\lambda) + 2 = 0$ and let $\Delta'(\mu') \geq 0$. Then $\Delta'(\lambda) > 0$ in any open interval $\mu' < \lambda < \mu_1'$ in which $\Delta(\lambda) < 2$.

Before proving Lemma 2.4, we may observe that it proves both the inequalities (2.4) and the assertion that the open intervals (2.6) are intervals of stability. In fact, we see from Lemma 2.1 that if $\lambda \leq \lambda^*$, then $\Delta(\lambda) > 2$. Among the infinitely many real zeros of the function $\Delta(\lambda) - 2$ there must be a smallest one which we call λ_0. We shall prove later (see Lemma 2.6) that $\Delta'(\lambda_0) < 0$. Therefore Lemma 2.4 shows that for $\lambda > \lambda_0$, $\Delta(\lambda)$ must be a decreasing function until $\Delta(\lambda) = -2$. This must actually happen for a certain $\lambda = \lambda_1' > \lambda_0$, since $\Delta(\lambda) + 2$ has infinitely many real zeros without a finite limit point. Now, either $\Delta'(\lambda_1') = 0$, or $\Delta'(\lambda_1') < 0$. If $\Delta'(\lambda_1') = 0$, λ_1' is a double root of $\Delta(\lambda) + 2 = 0$ and will be listed as λ_1' and as λ_2'. (According to Lemma 2.5 proved below, $\Delta(\lambda) + 2 = 0$ cannot have roots of multiplicity higher than two.) $\Delta'(\lambda_1') = 0$ implies (cf. Lemma 2.4) that $\Delta(\lambda)$ increases for $\lambda > \lambda_2' = \lambda_1'$ until it reaches the value 2. On the other hand, if $\Delta'(\lambda_1') < 0$, then $\Delta(\lambda) < -2$ for $\lambda_1' < \lambda < \lambda_2'$, where λ_2' is the smallest zero of $\Delta(\lambda) + 2$ which is $> \lambda_1'$. Since $\Delta(\lambda) < -2$ in the interval (λ_1', λ_2') this is an interval of instability for the solutions of (2.1) (cf. the stability test of Section 1.2). Now, $\Delta'(\lambda_2') \geq 0$, as can be seen from the fact that $\Delta(\lambda) < \Delta(\lambda_2')$ for all $\lambda < \lambda_2'$ and sufficiently close to λ_2'. Using Lemma 2.4 we may therefore conclude that $\Delta(\lambda)$ is an increasing function of λ in any interval $\lambda_2' < \lambda < \hat{\lambda}$ in which $\Delta(\lambda) < 2$. The largest interval of this kind is the interval (λ_2', λ_1), where λ_1 denotes the smallest root of $\Delta(\lambda) - 2$ which is $> \lambda_2'$. The stability test of Section 1.2 shows that this interval is an interval of stability for the solutions of (2.1).

Continuing in this manner we find that the inequalities (2.4) hold and that the open intervals (2.6) are the only intervals of stability for the solutions of (2.1).

We now prove Lemma 2.4. For this purpose, we introduce the following notations:

$$\frac{\partial}{\partial\lambda} y_1(x, \lambda) = z_1(x, \lambda), \qquad \frac{\partial}{\partial\lambda} y_2(x, \lambda) = z_2(x, \lambda),$$

$$\frac{\partial}{\partial\lambda} y_1'(x, \lambda) = z_1'(x, \lambda), \qquad \frac{\partial}{\partial\lambda} y_2'(x, \lambda) = z_2'(x, \lambda)$$

where obviously $z_1' = (\partial/\partial x)z_1$ and $z_2' = (\partial/\partial x)z_2$. Also, we shall write η_1, η_2, η_1', and η_2', respectively, for $y_1(\pi, \lambda)$, $y_2(\pi, \lambda)$, $y_1'(\pi, \lambda)$, and $y_2'(\pi, \lambda)$. As before, we shall write Δ for $\eta_1 + \eta_2'$ and Δ' for the derivative of Δ with respect to λ, that is:

$$\Delta' = z_1(\pi, \lambda) + z_2'(\pi, \lambda).$$

Next, we shall derive the formula

$$(2.13) \quad \Delta'(\lambda) = (\eta_1 - \eta_2') \int_0^\pi y_1(x) y_2(x)\, dx - \eta_2 \int_0^\pi y_1^2(x)\, dx + \eta_1' \int_0^\pi y_2^2(x)\, dx.$$

To prove (2.13), we differentiate (2.1) with respect to λ and we obtain (for $y = y_1$ and $y = y_2$, respectively)

$$\begin{aligned} z_1'' + (\lambda + Q)z_1 &= -y_1 \\ z_2'' + (\lambda + Q)z_2 &= -y_2. \end{aligned} \tag{2.14}$$

The general formula for the solution of an inhomogeneous linear differential equation of the second order in terms of the solution of the homogeneous equation yields

$$\begin{aligned} z_1(x) &= y_1(x) \int_0^x y_2(t)y_1(t)\,dt - y_2(x) \int_0^x y_1^2(t)\,dt \\ z_1'(x) &= y_1'(x) \int_0^x y_2(t)y_1(t)\,dt - y_2'(x) \int_0^x y_1^2(t)\,dt \\ z_2(x) &= y_1(x) \int_0^x y_2^2(t)\,dt - y_2(x) \int_0^x y_1(t)y_2(t)\,dt \\ z_2'(x) &= y_1'(x) \int_0^x y_2^2(t)\,dt - y_2'(x) \int_0^x y_1(t)y_2(t)\,dt. \end{aligned} \tag{2.15}$$

The functions z_1 and z_2 in (2.15) are those solutions of (2.14) which satisfy the initial conditions $z_1(0) = z_1'(0) = 0$ and $z_2(0) = z_2'(0) = 0$. Since $z_1 = (\partial/\partial\lambda)y_1$ etc., and since the initial conditions for y_1 and y_2 are independent of λ, the solutions (2.15) are the correct ones. Equation (2.13) follows immediately from (2.15) by putting $x = \pi$.

Since the Wronskian $y_1y_2' - y_2y_1' = 1$ for all x, we find for $x = \pi$ that

$$\eta_1\eta_2' - \eta_2\eta_1' = 1. \tag{2.16}$$

Hence

$$\Delta^2 - 4 = (\eta_1 + \eta_2')^2 - 4(\eta_1\eta_2' - \eta_2\eta_1') = (\eta_1 - \eta_2')^2 + 4\eta_1'\eta_2.$$

Putting $\operatorname{sgn} \eta_1' = +1$ if $\eta_1' > 0$, $\operatorname{sgn} \eta_1' = -1$ if $\eta_1' < 0$ and $\operatorname{sgn} \eta_1' = 0$ if $\eta_1' = 0$, and assuming that $\eta_1' \neq 0$, we find from (2.13):

$$\Delta'(\lambda) = \operatorname{sgn} \eta_1' \left\{\int_0^\pi \left(\sqrt{|\eta_1'|}\, y_2 + \operatorname{sgn} \eta_1' \frac{\eta_1 - \eta_2'}{2\sqrt{|\eta_1'|}}\, y_1\right)^2 dx - \frac{\Delta^2 - 4}{4|\eta_1'|} \int_0^\pi y_1^2\,dx\right\}. \tag{2.17}$$

Equation (2.17) shows that $\Delta'(\lambda)$ has the same sign as η_1' in any interval in which $\eta_1' \neq 0$ and $\Delta^2 \leq 4$. Consider now a value μ of λ such that

$\Delta(\mu) = 2$ and $\Delta'(\mu) \leq 0$. We wish to establish the fact that, for a sufficiently small δ, $\Delta(\lambda)$ is decreasing in the interval $\mu < \lambda < \mu + \delta$. If $\Delta'(\mu) < 0$, this is obvious. Assume now that $\Delta(\mu) = 2$ and $\Delta'(\mu) = 0$. In this case, according to (2.17), we must have $\eta_1'(\mu) = 0$. Since we also have

$$\Delta^2 - 4 = (\eta_1 - \eta_2')^2 + 4\eta_1'\eta_2 = 0,$$

we find that $\eta_1(\mu) - \eta_2'(\mu) = 0$ and, since $\eta_1\eta_2' - \eta_2\eta_1' = 1$, we have $\eta_1(\mu) = \eta_2'(\mu) = 1$. Then (2.13) reduces to

$$\Delta'(\mu) = -\eta_2 \int_0^\pi y_1^2(x)\,dx,$$

and therefore $\Delta'(\mu) = 0$ implies $\eta_2(\mu) = 0$.

Now we shall compute

$$\Delta''(\lambda) = d\Delta'(\lambda)/d\lambda$$

for $\lambda = \mu$, where μ is such that

$$\eta_1'(\mu) = \eta_2(\mu) = 0, \quad \eta_1(\mu) = \eta_2'(\mu) = 1. \tag{2.18}$$

We shall do this by differentiating (2.13) with respect to λ and by using (2.15) for $x = \pi$ in order to obtain the derivatives with respect to λ of $\eta_1'(\lambda)$, etc., for $\lambda = \mu$. A straightforward computation shows that, if (2.18) holds,

$$\Delta''(\mu) = 2\left\{\int_0^\pi y_1(x)y_2(x)\,dx\right\}^2 - 2\int_0^\pi y_1^2(x)\,dx \int_0^\pi y_2^2(x)\,dx. \tag{2.19}$$

Since $y_1(x)$ and $y_2(x)$ are linearly independent functions, we see from an application of the Schwartz inequality to (2.19) that

$$\Delta''(\mu) < 0. \tag{2.20}$$

Thus $\Delta'(\lambda)$ is again found to be decreasing in an interval $\mu < \lambda < \mu + \delta$. Assume now that Lemma 2.4 is false. Then there would exist a smallest number $\mu^* > \mu$ such that $\Delta'(\lambda) < 0$ for $\mu < \lambda < \mu^*$ but $\Delta'(\mu^*) = 0$ although $\Delta(\mu^*) > -2$. We would then have

$$\Delta^2(\mu^*) - 4 = (\eta_1 - \eta_2')^2 + 4\eta_1'\eta_2 < 0 \tag{2.21}$$

and therefore $\eta_1'\eta_2 < 0$ for $\lambda = \mu^*$. But then, $\eta_1'(\mu^*) \neq 0$ and so, according to (2.17), $\Delta'(\mu^*) \neq 0$, which produces a contradiction. This proves Lemma 2.4 in the case where $\Delta(\lambda) = 2$.

If $\Delta(\lambda) = -2$, the proof is almost literally the same. Incidentally, our proof of Lemma 2.4 shows that the following is true:

Lemma 2.5. The roots of the equation

$$\Delta^2(\lambda) - 4 = 0$$

are either simple or double roots. If, for a particular value of $\lambda = \mu$,

$$\Delta^2(\mu) = 4, \quad \Delta'(\mu) = 0, \tag{2.22}$$

then $\Delta''(\mu) < 0$ if $\Delta(\mu) = 2$ and $\Delta''(\mu) > 0$ if $\Delta = -2$. Necessary and sufficient conditions for $\Delta^2(\mu) - 4$ and $\Delta'(\mu)$ to vanish simultaneously are

$$\eta_1(\mu) - \eta_2'(\mu) = \eta_1'(\mu) = \eta_2(\mu) = 0. \tag{2.23}$$

In order to complete the proof of Theorem 2.1 we need

Lemma 2.6. Let λ_0 be the smallest root of the equation $\Delta^2(\lambda) - 4 = 0$. Then λ_0 is a simple root and $\Delta'(\lambda_0) < 0$.

Proof. We know from Lemma 2.1 that $\Delta(\lambda) > 2$ for $\lambda < \lambda_0$. Therefore $\lambda = \lambda_0$ cannot be a maximum of $\Delta(\lambda)$ if $\Delta(\lambda_0) = 2$. But Lemma 2.5 shows that $\Delta(\lambda)$ would have a maximum at $\lambda = \lambda_0$ if $\Delta'(\lambda_0) = 0$. This is a contradiction, and therefore Lemma 2.6 is true.

A comparison of Lemma 2.5 and of Floquet's theorem in Chapter I shows immediately that we can supplement Theorem 2.1 by the following

Corollary 2.1. *Hill's equation* (2.1) *has two linearly independent periodic solutions of period* π *or* 2π *if and only if the equation* $\Delta^2(\lambda) - 4 = 0$ *has a double root.*

The stability test at the end of Section 1.2, Corollary 2.1, and Lemma 2.5 show that the solutions of (2.1) are stable for $\lambda = \lambda_{2n+1}$ (or $\lambda = \lambda'_{2n+1}$) if and only if λ_{2n+1} (or λ'_{2n+1}) is a double root of $\Delta^2(\lambda) - 4$, that is, if and only if $\lambda_{2n+1} = \lambda_{2n+2}$ (or $\lambda'_{2n+1} = \lambda'_{2n+2}$). This proves Theorem 2.1 completely.

2.2. *Analytic properties of the discriminant*

The function $\Delta(\lambda)$ as defined by (2.12) will be called the *discriminant* of Hill's equation (2.1). In this section, we shall prove several results about $\Delta(\lambda)$, the first of which we used already in Section 2.1. Unless otherwise stated, we shall assume that $Q(x)$ in (2.1) is of bounded variation. We have:

Theorem 2.2. *The function*

$$\Delta(\lambda) = y_1(\pi, \lambda) + y_2'(\pi, \lambda)$$

is an entire analytic function of the complex variable λ. Its order of growth for $|\lambda| \to \infty$ is exactly $\frac{1}{2}$; i.e., there exists a positive constant M such that

$$|\Delta(\lambda)| \exp(-M\sqrt{|\lambda|}) \tag{2.24}$$

is bounded for all λ and a positive constant m such that, λ real and $\lambda \to -\infty$ implies

$$|\Delta(\lambda)| \exp(-m\sqrt{|\lambda|}) \to \infty. \tag{2.25}$$

Corollary 2.2. *The functions $\Delta(\lambda) + 2$ and $\Delta(\lambda) - 2$ have infinitely many zeros.*

Note that Corollary 2.2 follows immediately from Theorem 2.2, which permits us to conclude that $\Delta(\lambda) + 2$ and $\Delta(\lambda) - 2$ are functions of order of growth $\frac{1}{2}$. According to a well-known theorem on entire functions [see Levin (1964) or Titchmarsh (1938)] any entire function of order of growth $\frac{1}{2}$ has infinitely many zeros.

In order to prove Theorem 2.2, we apply Picard's method of iteration to the differential equation (2.1). Let $\omega = \sqrt{\lambda}$ and let

$$u_0 = \cos \omega x, \quad v_0 = \frac{\sin \omega x}{\omega}$$

and define u_n and v_n recursively for $n = 1, 2, \ldots$ by

$$u_n(x, \omega) = -\frac{1}{\omega}\int_0^x \sin \omega(x - \xi)Q(\xi)u_{n-1}(\xi)\, d\xi \tag{2.26}$$

$$v_n(x, \omega) = -\frac{1}{\omega}\int_0^x \sin \omega(x - \xi)Q(\xi)v_{n-1}(\xi)\, d\xi. \tag{2.27}$$

Then

$$y_1(x, \omega^2) = \sum_{n=0}^{\infty} u_n(x, \omega) \tag{2.28}$$

(2.29) $$y_2(x, \omega^2) = \sum_{n=0}^{\infty} v_n(x, \omega)$$

and

(2.30) $$\Delta(\lambda) = \Delta(\omega^2) = \sum_{n=0}^{\infty} [u_n(\pi, \omega) + v_n'(\pi, \omega)],$$

where $v_0'(x, \omega) = \cos \omega x$ and

(2.31) $$v_n'(x, \omega) = -\int_0^x \cos \omega(x - \xi)Q(\xi)v_{n-1}(\xi, \omega)\, d\xi.$$

It is easy to see that for all real values of $x \geq 0$:

$$|u_0| \leq e^{|\omega|x}, \quad |v_0| \leq xe^{|\omega|x}.$$

Now let M^* be a positive constant such that, for all real values of x, $|Q(x)| \leq M^*$. Then we find by induction from (2.26), (2.27), and from

$$|\sin \omega(x - \xi)| \leq |\omega|(x - \xi)e^{|\omega|(x-\xi)} \qquad (0 \leq \xi \leq x)$$

that

(2.32) $$|u_n(x, \omega)| \leq e^{|\omega|x}(M^*x^2)^n/(2n)!$$

(2.33) $$|v_n(x, \omega)| \leq xe^{|\omega|x}(M^*x^2)^n/(2n + 1)!.$$

Equations (2.33) and (2.31) show that

(2.34) $$|v_n'(x, \omega)| \leq e^{|\omega|x}(M^*x^2)^n/(2n)!$$

and we therefore conclude from (2.30) that

(2.35) $$|\Delta(\omega^2)| \leq 2e^{|\omega|\pi} \cosh(\sqrt{M^*}\pi).$$

This proves that the expression (2.24) is bounded if we choose $M = \pi$. In order to complete the proof of Theorem 2.2, we have to prove (2.25). For this purpose, we may assume that $Q(x) \leq -1$ for all x. Otherwise, we could replace Q by $(Q - M^* - 1)$ and λ by $(\lambda + M^* + 1)$ without changing the differential equation for y. Putting $\sqrt{\lambda} = i\phi = \omega$, where ϕ is real and positive and $\phi \to \infty$ as $\lambda \to -\infty$, we find

$$u_0 \geq \tfrac{1}{2}e^{\phi x} \geq \tfrac{1}{2}e^{\phi x/2}$$

$$\frac{\sin \omega(x - \xi)}{\omega} = \frac{\sinh \phi(x - \xi)}{\phi} =$$

$$\frac{1}{2}\int_{\xi - x}^{x - \xi} e^{\phi s}\, ds \geq \frac{1}{2}\int_{(x-\xi)/2}^{x-\xi} e^{\phi s}\, ds \geq \frac{1}{4}(x - \xi)\, e^{\phi(x-\xi)/2}.$$

From the preceding inequalities and from (2.26) we find by induction with respect to n that

$$u_n(x, i\phi) \geq \tfrac{1}{2}e^{\phi x/2}(x/2)^{2n}/(2n)!. \tag{2.36}$$

By an even simpler argument, we can prove that $v_n'(x, i\phi) \geq 0$, and therefore we find from (2.36) and (2.30) that

$$\Delta(\lambda) = \Delta(-\phi^2) \geq \sum_{n=0}^{\infty} u_n(\pi, i\phi) \geq \tfrac{1}{2}e^{\phi\pi/2} \cosh(\pi/2). \tag{2.37}$$

This proves that (2.25) is true for any m between 0 and $\pi/2$.

Obviously, all the functions u_n and v_n' are entire analytic functions of λ and since we have shown that their sums converge uniformly in any finite part of the λ plane, it follows that $\Delta(\lambda) = y_1(\pi, \lambda) + y_2'(\pi, \lambda)$ is also an entire function of λ. The inequalities (2.35) and (2.37) establish the truth of our assertion about the order of growth of $\Delta(\lambda)$ and this completes the proof of Theorem 2.2.

For later purposes we note here a result which can be proved by exactly the same method as Theorem 2.2.

Theorem 2.3. *Let $y(x, \lambda)$ be any real solution of Hill's equation* (2.1) *with initial conditions independent of λ. Let x be fixed and real, and consider $y(x, \lambda)$ and $y'(x, \lambda)$ as functions of λ. Then the order of growth of these two functions of λ is at most $\tfrac{1}{2}$.*

The following theorem gives an idea about the asymptotic behavior of $\Delta(\lambda)$ for positive values of λ.

Theorem 2.4. *The absolute value of the function*

$$\sqrt{\lambda}[\Delta(\lambda) - 2\cos \pi\sqrt{\lambda}]$$

tends to zero as $\lambda \to +\infty$ if *$Q(x)$ is normalized so that $\int_0^\pi Q\,dx = 0$.*

For all complex values of $\sqrt{\lambda}$ with nonnegative real part.

$$|\Delta(\lambda) - 2\cos \pi\sqrt{\lambda}| \leq 2e^{\pi\omega_2}[\exp(\pi M^*/|\sqrt{\lambda}|) - 1], \tag{2.38}$$

where ω_2 denotes the absolute value of the imaginary part of $\sqrt{\lambda}$ and where M^* is such that $|Q(x)| \leq M^*$.

Proof. It is clear that the second statement of Theorem 2.4 implies the boundedness of

$$\sqrt{\lambda}E = \sqrt{\lambda}|\Delta(\lambda) - 2\cos\pi\sqrt{\lambda}|;$$

because for small values of $\sqrt{\lambda}$ (e.g., for $|\lambda| \leq 1$)E is bounded anyway, and for larger values of $|\sqrt{\lambda}| = t$ we can use the estimate

$$\exp(\pi M^*/t) - 1 = \int_0^{\pi M^*/t} e^s\,ds \leq (\pi M^*/t)\exp(\pi M^*/t) \leq (\pi M^*/t)e^{\pi M^*}$$

Therefore, if λ is positive and real (*i.e.*, $\omega_2 = 0$), the boundedness of $\sqrt{\lambda}E$ follows from (2.38).

In order to prove (2.38), we use induction on $n = 1, 2, 3, \ldots$ to prove that (with $\omega = \sqrt{\lambda}$), for $x \geq 0$

$$|u_n(x, \omega)| \leq e^{\omega_2 x}M^{*n}x^n|\omega|^{-n}/n!, \qquad |v_n'(x, \omega)| \leq e^{\omega_2 x}M^{*n}x^n|\omega|^{-n}/n!.$$

These inequalities can be proved by using the procedure which led to the proof of the first part of Theorem 2.2. We merely have to replace the estimates used there by the inequalities

$$|u_0| \leq e^{\omega_2 x}, \quad |v_0| \leq xe^{\omega_2 x} \tag{2.39}$$

and

$$|\cos\sqrt{\lambda}(x - \xi)| \leq \exp\omega_2(x - \xi)$$
$$|\sin\sqrt{\lambda}(x - \xi)| \leq \exp\omega_2(x - \xi)$$

which are proved easily.

In order to show that $\sqrt{\lambda}E \to 0$ as $\lambda \to +\infty$, we have merely to prove that

$$\lim_{\omega\to\infty}\omega|u_1(\pi, \omega) + v_1'(\pi, \omega)| = 0$$

since, for $n > 1$, $u_n(\pi, \omega)$ and $v_n'(\pi, \omega)$ tend to zero at least as ω^{-2} for $\omega \to \infty$. Now we have, because of the normalizing condition for Q,

$$\omega\{u_1(\pi, \omega) + v_1'(\pi, \omega)\} = -\int_0^\pi \sin(\omega\pi)Q(\xi)\,d\xi = 0.$$

This completes the proof of Theorem 2.4.

Obviously, Theorem 2.4 is true for all bounded, square-integrable functions $Q(x)$. In order to obtain better results pertaining to the asymptotic behavior of $\Delta(\lambda)$ for $\lambda > 0$, $\lambda \to \infty$, we shall assume that $Q(x)$ can be expanded in a Fourier series

$$Q(x) = \sum_{n=-\infty}^{+\infty}{}' g_n e^{2inx}, \tag{2.40}$$

where the prime at the summation symbol indicates that the sum is extended over values of $n \neq 0$ only. This does not impose any restriction on $Q(x)$, since a constant term in (2.40) could be combined with the parameter λ in Hill's equation (2.1). Equivalently, we could say that $Q(x)$ has been normalized so that

$$\int_0^\pi Q(x)\,dx = 0. \tag{2.41}$$

Since $Q(x)$ is real, the constants g_n in (2.40) satisfy the condition

$$g_{-n} = \bar{g}_n \tag{2.41*}$$

for all n, where a bar denotes the complex conjugate quantity. We shall always assume that

$$\sum_{n=-\infty}^{+\infty} |g_n| < \infty \tag{2.42}$$

and in many cases that

$$\lim_{n \to \pm\infty} n^2 g_n = 0 \tag{2.43}$$

(This will be true if $Q(x)$ has a continuous second derivative.)

If we put

$$\Delta_n(\lambda) = u_n(\pi, \sqrt{\lambda}) + v_n'(\pi, \sqrt{\lambda}), \tag{2.44}$$

where u_n and v_n are defined by (2.26) and (2.27), then

$$\Delta(\lambda) = \sum_{n=0}^{\infty} \Delta_n(\lambda).$$

Obviously, $\Delta_n(\lambda)$ is a homogeneous form of degree n in the infinitely many variables g_n, that is,

$$\Delta_n(\lambda) = \sum_{l_1,\ldots,l_n=-\infty}^{\infty} c(l_1, \ldots, l_n) g_{l_1} \cdots g_{l_n} \tag{2.45}$$

where we may sum without restrictions if we assume that $g_0 = 0$. Now we shall prove

Theorem 2.5. *Let $\omega = \sqrt{\lambda}$. Then*

$$c(l_1, \ldots, l_n) = A(\omega) \cos \pi\omega + B(\omega) \frac{\sin \pi\omega}{\omega},$$

where $A(\omega)$ and $B(\omega)$ are even rational functions of ω such that

(1) *In each of the functions $A(\omega)$ and $B(\omega)/\omega$ the degree of the denominator exceeds the degree of the numerator by at least n.*

(2) *The poles of $A(\omega)$ and $B(\omega)$ are at some of the points $\omega = 0$ and*

$$\omega = \pm (l_r + l_{r+1} + \cdots + l_s),$$

where $1 \leq r \leq s \leq n$.

Proof of Theorem 2.5. We shall use the expression for Δ_n in terms of $u_n(\pi, \lambda)$ and $v_n'(\pi, \lambda)$. However, the expressions of u_n and v_n' as given by (2.26) and (2.31) involve n integrations, and direct evaluation of these integrals leads to results which are cumbersome in form. We shall therefore make use of the fact that the integrals in (2.26) and (2.31) are convolution integrals, the Laplace transforms of which can be easily computed. We find for the coefficient of

$$g_{l_1} g_{l_2} \cdots g_{l_n}$$

(the factors in this order) in the Laplace integral

$$\int_0^\infty e^{-px}[u_n(x, \lambda) + v_n'(x, \lambda)]\, dx$$

the expression

$$(2.46) \quad (-1)^n k(l_1, \ldots, l_n, p) = \frac{2p - 2i \sum_{v=1}^{n} l_v}{[\omega^2 + p^2] \prod_{v=1}^{n} [\omega^2 + (p - 2il_1 - \ldots - 2il_v)^2]}.$$

By an application of the inversion formula for the Laplace transformation we find that $c(l_1, \ldots, l_n)$ is the sum of the residues of

$$U = e^{\pi p} k(l_1, \ldots, l_n, p).$$

Obviously, the poles of U are located at

$$p = \pm i\omega, \pm i\omega + 2il_1, \pm i\omega + 2il_1 + 2il_2, \ldots$$

The rational function k of p in (2.46) has a denominator of degree $(2n + 2)$ and a numerator of degree 1. The zeros of the denominator will have a multiplicity m which does not exceed $(n + 1)$ if we treat ω as a variable. Let ζ be one of the poles of k, and put

$$U = e^{\pi p}k^*(p)(p - \zeta)^{-m}$$

where k^* is a rational function of p, regular at $p = \zeta$, in which the degree of the denominator exceeds that of the numerator by $2n + 1 - m$. Now we obtain the residue of U at $p = \zeta$ by differentiating $(\exp \pi p)k^*$ exactly $(m - 1)$ times with respect to p and then putting $p = \zeta$. Now all derivatives of $(\exp \pi p)k^*$ are of the form $k^{**} \exp \pi p$, where k^{**} is rational in p and such that the degree of the denominator exceeds that of the numerator by at least $2n + 1 - m$. Since ζ is linear in ω, statement (1) of Theorem 2.5 has been proved. (Observe that A and B *must* be even functions of ω.)

Since the product $g_{l_1} \cdots g_{l_n}$ does not change if we permute the variables $g_{l_1}, \cdots, g_{l_n}$, we find that $\Delta_n(\lambda)$ can also be written in the form

$$\Delta_n(\lambda) = \sum_{l_1 \leq l_2 \leq, \ldots \leq l_n} \gamma(l_1, \ldots, l_n) g_{l_1} \cdots g_{l_n} \tag{2.47}$$

where $-\infty < l_1 \leq l_2 \cdots \leq l_n < \infty$. The coefficients γ in (2.47) are defined by the relation

$$\gamma(l_1, \ldots, l_n) = \sum c(K_1, \ldots, K_n), \tag{2.48}$$

where the sum in (2.48) is to be extended over all distinct sets of integers $K_1, \ldots, K_n$ which can be made to coincide with the set $l_1, \ldots, l_n$ by a suitable permutation of the indices $1, \ldots, n$. Concerning the coefficients γ we have

Theorem 2.6. *The coefficients γ defined by* (2.48) *satisfy the relations*

$$\gamma(l_1, \ldots, l_n) = 0$$

whenever

$$l_1 + \cdots + l_n \neq 0.$$

Proof. If we replace $Q(x)$ in (2.1) by $Q(x + \phi)$, where ϕ is a constant, the resulting differential equation will have the same discriminant $\Delta(\lambda)$

as the original one. In fact, it is clear that if $Q(x)$ is replaced by $Q(x + \phi)$, Hill's equation continues to have periodic solutions for the same values of λ and in each case their number is unaffected by the change. Since $\Delta(\lambda) - 2$ is an analytic function of λ whose order of growth is $\frac{1}{2}$, $\Delta(\lambda) - 2$ is determined by its zeros up to a multiplicative constant. Assume now that $\Delta(\lambda) - 2$ belongs to the function $Q(x)$ and that $\Delta^*(\lambda) - 2$ belongs to the function $Q(x + \phi)$. Then

$$\Delta(\lambda) - 2 = C[\Delta^*(\lambda) - 2],$$

where C is a constant. On the other hand, $\Delta(\lambda) + 2$ and $\Delta^*(\lambda) + 2$ also have the same zeros (with the same multiplicity), and therefore

$$\Delta(\lambda) + 2 = C'[\Delta^*(\lambda) + 2].$$

By subtracting the first of these equations from the second we find that

$$(C' - C)\Delta^*(\lambda) + 2C' + 2C = 4.$$

Therefore $\Delta^*(\lambda)$ would be a constant unless $C' = C = 1$. However, Theorem 2.4 shows that $\Delta^*(\lambda)$ can never be a constant. Hence $\Delta(\lambda) = \Delta^*(\lambda)$ for all λ.

It is obvious that $\Delta^*(\lambda)$ arises from $\Delta(\lambda)$ by substituting $g_n e^{2in\phi}$ for g_n, and therefore the coefficient of $g_{l_1} \cdots g_{l_n}$ in Δ^* will be

$$\gamma(l_1, \ldots, l_n)e^{2iL\phi},$$

where $L = l_1 + \cdots + l_n$. Assume now that all of the g_n are zero except for $g_{l_1}, g_{l_2}, \ldots, g_{l_n}$. The same argument which we used in showing that $\Delta(\lambda)$ is an analytic function of λ (Theorem 2.2) can be used to show that Δ depends analytically on $g_{l_1}, \ldots, g_{ln}$ and the same is true for Δ^*. Hence, $\Delta(\lambda) = \Delta^*(\lambda)$ implies that the coefficients of $g_{l_1} \cdots g_{l_n}$ in both functions must be the same, so that for all real values of ϕ:

$$\gamma(l_1, \ldots, l_n) = e^{2iL\phi}\gamma(l_1, \ldots, l_n).$$

If $L \neq 0$, this is possible only if $\gamma = 0$, as stated in Theorem 2.6. We may note here

Corollary 2.6. *The first terms in the expansion* (2.44) *are*

$$\Delta_0(\lambda) = 2\cos \pi\sqrt{\lambda}, \quad \Delta_1(\lambda) = 0$$

$$\Delta_2(\lambda) = \frac{\pi \sin \pi\sqrt{\lambda}}{2\sqrt{\lambda}} \sum_{n=1}^{\infty} \frac{|g_n|^2}{\lambda - n^2}$$

Of these relations, the first one is trivial, the second one follows from Theorem 2.6, and the third one is a result of a somewhat tedious computation.

2.3. *Infinite determinants*

Hill (1886) used infinite determinants for the investigation of the characteristic values of λ in (2.1). Whittaker and Watson (1927) showed that the value of Hill's determinant can be expressed in terms of $\Delta(\lambda)$. In this section we shall reproduce the results of Hill and Watson and supplement them by some rclations of the type obtained by Whittaker and Watson.

We shall write a determinant in the form

$$\|a_{n,m}\|_k^l$$

where n and m vary over all integers from k to l. In particular, we shall consider the determinants where $k = -\infty$ and $l = \infty$ or where $k = 0$ and $l = \infty$. These we shall call, respectively, two-sided infinite and one-sided infinite determinants.

We shall always use the first subscript n to denote the rows and the second subscript m to denote the columns of the determinant.

We shall say that the infinite determinants

$$\|a_{n,m}\|_0^\infty, \quad \|a_{n,m}\|_{-\infty}^\infty$$

exist or converge if the limits

$$\lim_{l\to\infty} \|a_{n,m}\|_0^l, \quad \lim_{l\to\infty} \|a_{n,m}\|_{-l}^l$$

exist. The value of the limit is then called the value of the determinant. We shall not be concerned here with a general theory of infinite determinants; instead, we shall introduce a special class of infinite determinants and indicate briefly the proof of a few theorems which hold for this class.

We shall say that a determinant is of *Hill's type* if it satisfies the condition

$$\sum_{n,m} |a_{n,m} - \delta_{n,m}| < \infty \tag{2.49}$$

where $\delta_{n,m} = 1$ for $n = m$ and $\delta_{n,m} = 0$ otherwise, and where the sum in (2.49) is to be taken over all values of n and m. Obviously, every finite determinant is of Hill's type. We shall show now:

Theorem 2.7. *An infinite determinant of Hill's type converges.*

It suffices to prove Theorem 2.7 in the case of a determinant $\|a_{n,m}\|_0^\infty$. According to a theorem due to Hadamard [see Hardy, Littlewood, and Polya (1934)] the absolute value of the square of any finite determinant

$$\|a_{n,m}\|_k^l$$

does not exceed the value of the product

$$\prod_{n=k}^{l}\left(\sum_{m=k}^{l}|a_{n,m}|^2\right)$$

From this we derive the following

Lemma 2.7. Let $\|a_{n,m}\|_0^\infty$ be a determinant of Hill's type. Let

$$a'_{n,m} = a_{n,m} \quad \text{if} \quad n \neq m$$
$$a'_{n,n} = a_{n,n} \quad \text{if} \quad |a_{n,n}| \geq 1$$
$$a'_{n,n} = 1 \quad \text{if} \quad |a_{n,n}| < 1,$$

and let

$$H = \left\{\prod_{n=0}^{\infty}\left(\sum_{m=0}^{\infty}|a'_{n,m}|^2\right)\right\}^{1/2}$$

Then $H < \infty$, and the absolute value of any finite subdeterminant of $\|a_{n,m}\|_0^\infty$ is at most equal to H.

Proof. The only difficulty is to show that $H < \infty$. For this purpose, let

$$\varepsilon_n = \sum_m{}' |a_{n,m}|,$$

where the sum is taken over all $m \neq n$. Also, let

$$\gamma_n = |a'_{n,n}| - 1.$$

It follows from (2.49) that

$$\sum_{n=0}^{\infty} \varepsilon_n < \infty, \quad \sum_{n=0}^{\infty} \gamma_n < \infty.$$

Finally, let

$$p_n = \left\{ \sum_{m=0}^{\infty} |a'_{n,m}|^2 - 1 \right\}^{1/2}.$$

Then

$$p_n^{\,2} \leq |a'_{n,n}|^2 - 1 + \left(\sum_m{}' \, |a_{n,m}| \right)^2 =$$

$$|a'_{n,n}|^2 - 1 + \varepsilon_n^{\,2} \leq A|a_{n,n} - 1| + \varepsilon_n^{\,2},$$

where $A = \max |a_{n,n}| + 1$.

In view of (2.49),

$$\sum_{n=0}^{\infty} |a_{n,n} - 1| < \infty.$$

Also, we know that the convergence of a sum of positive terms implies the convergence of the sum of the squares of the terms. Hence

$$\sum_{n=0}^{\infty} p_n^{\,2} < \infty$$

and

$$H = \left[\prod_{n=0}^{\infty} (1 + p_n^{\,2}) \right]^{1/2} < \infty.$$

This proves Lemma 2.7.

Now we can prove Theorem 2.7. We can expand

$$\|a_{n,m}\|_0^{l+1} = D_{l+1}$$

in terms of the elements of the last row and their subdeterminants which can be majorized by the quantity H defined in Lemma 2.7. The result is

$$D_{l+1} = a_{l+1,l+1} D_l + \theta_l H,$$

where $|\theta_l| \leq \varepsilon_{l+1}$. Since $|D_l| \leq H$, it follows that

$$|D_{l+1} - D_l| \leq H|a_{l+1,l+1} - 1| + H\varepsilon_{l+1}$$

$$= H \sum_{m=0}^{\infty} |a_{l+1,m} - \delta_{l+1,m}|.$$

Therefore

$$\sum_{l=1}^{\infty} |D_{l+1} - D_l| < \infty$$

and

$$\lim_{l \to \infty} D_{l+1} = D$$

exists. This proves Theorem 2.7.

The last result about infinite determinants which we need is

Theorem 2.8. *Let $\|a_{n,m}\|$ be an infinite determinant of Hill's type, and assume that there exist numbers x_m not all of which vanish such that $|x_m| \leq M$ (M fixed) for all m and*

$$\sum_m a_{n,m} x_m = 0$$

for all n. Then

$$\|a_{n,m}\| = 0.$$

Proof. Since the set of subdeterminants of $\|a_{n,m}\|$ is bounded, it follows that for each n

$$\|a_{n,m}\| x_n = 0.$$

This relation is obtained in the same manner in which the corresponding relation is obtained in the case of a finite system of linear equations. The inequality (2.49) and the condition $|x_m| \leq M$ guarantee the absolute convergence of all sums involved. This proves Theorem 2.8.

We shall now express the discriminant $\Delta(\lambda)$ of Hill's equation in terms of an infinite determinant. For this purpose we shall write (2.1) in the form

$$y'' + \left(\sum_{n=-\infty}^{\infty} g_n e^{2inx}\right) y = 0, \tag{2.50}$$

where $\lambda = g_0$ and $Q(x)$ is given by (2.40). We know from Floquet's theorem that (2.50) has a solution $\neq 0$ of the type

$$y = e^{i\alpha x} p(x), \tag{2.51}$$

where $p(x)$ is a function of period π and where

$$2 \cos \pi\alpha = y_1(\pi) + y_2'(\pi).$$

If $Q(x)$ is sufficiently smooth, e.g., if $\sum |g_n| < \infty$, the function y in (2.51) can be expanded in a twice termwise-differentiable series

$$y(x) = \sum_{n=-\infty}^{\infty} p_n e^{i(\alpha+2n)x}, \tag{2.52}$$

and the left-hand side of (2.50) takes the form

$$\sum_{n=-\infty}^{\infty} C_n e^{i(\alpha+2n)x}. \tag{2.53}$$

Since (2.53) must vanish identically, we have $C_n = 0$ for all n. If we write C_n explicitly in terms of the p_n and g_n, we have, for $-\infty < n < \infty$,

$$\sum_{m=-\infty}^{\infty} [g_{n-m} - (\alpha + 2n)^2 \delta_{n,m}] p_m = 0, \tag{2.54}$$

or, after multiplication by $[g_0 - (\alpha + 2n)^2]^{-1}$:

$$\sum_{m=-\infty}^{\infty} \left[\frac{g_{n-m}}{\lambda - (\alpha + 2n)^2} + \delta_{n,m}\right] p_m = 0, \tag{2.55}$$

where we have replaced g_0 by λ and where, just as in (2.41*)

$$g_{n-m} = \bar{g}_{m-n}, \quad g_0 = 0. \tag{2.56}$$

Obviously, the determinant

$$D(\alpha, \lambda) = \left\|\frac{g_{n-m}}{\lambda - (\alpha + 2n)^2} + \delta_{n,m}\right\|_{-\infty}^{\infty} \tag{2.57}$$

converges if $\sum |g_n| < \infty$, except for such values of λ and α for which one of the denominators $\lambda - (\alpha + 2n)^2$ vanishes.

It is easy to see that

Lemma 2.8. $D(\alpha, \lambda)$ regarded as a function of α is single valued and analytic for all values of α other than the values

$$\alpha = \pm\sqrt{\lambda} - 2n, \qquad n = 0, \pm 1, +2, \ldots, \tag{2.58}$$

at which the function may have poles. If $\lambda \neq 0$, these poles are (at most) of order one. $D(\alpha, \lambda)$ is periodic (in α) with period 2 and for $\alpha \to i\infty$, $D(\alpha, \lambda) \to 1$.

The proof of Lemma 2.8 is mostly routine except for the statement about the periodicity of $D(\alpha, \lambda)$. This follows from the remark that D remains unchanged if we replace α by $\alpha + 2$ and at the same time replace n by $n - 1$ and m by $m - 1$. Since both n and m run from $-\infty$

to $+\infty$, the same is true for $n - 1$ and $m - 1$, and therefore D does not change if we replace α by $\alpha + 2$. The rest of the proof of Lemma 2.8 is left to the reader.

Since the residues of

$$\frac{g_{n-m}}{\lambda - (\alpha + 2n)^2}$$

at the values $\alpha = \sqrt{\lambda} - 2n$ and $\alpha = -\sqrt{\lambda} - 2n$ add up to zero, it follows from the periodicity of $D(\alpha, \lambda)$ that (for $\lambda \neq 0$) all of its residues have the same value K for $\alpha = \sqrt{\lambda} - 2n$ (independent of n) and the value $-K$ for $\alpha = -\sqrt{\lambda} - 2n$. Therefore,

$$E(\alpha) = D(\alpha, \lambda) - \frac{\pi K}{2}\left\{\operatorname{ctg}\frac{\pi}{2}(\alpha - \sqrt{\lambda}) - \operatorname{ctg}\frac{\pi}{2}(\alpha + \sqrt{\lambda})\right\} \tag{2.59}$$

is an entire function of α with period 2. Since $E(\alpha)$ is bounded in the strip

$$-1 \leq \operatorname{Re} \alpha \leq 1,$$

$E(\alpha)$ is a constant E. We wish to determine E and K. By letting $\alpha \to i\infty$, we find from Lemma 2.8 that $E = 1$. We cannot determine K explicitly, but we can express it in terms of $D(0, \lambda)$ by putting $\alpha = 0$. The result is

$$K = \frac{1}{\pi}[1 - D(0, \lambda)] \operatorname{tg}\frac{\pi}{2}\sqrt{\lambda}. \tag{2.60}$$

We can now apply Theorem 2.8 to (2.55). The infinitely many equations (2.55) are not always the equivalent of (2.54), since α may have one of the exceptional values (2.58). However, we may multiply (2.55) by

$$\left(1 + \frac{\alpha - \sqrt{\lambda}}{2n}\right)\left(1 + \frac{\alpha + \sqrt{\lambda}}{2n}\right) = \frac{(2n + \alpha)^2 - \lambda}{4n^2}$$

for $n \neq 0$ and by $\alpha^2 - \lambda$ for $n = 0$. The determinant of the resulting system must vanish if not all of the p_m vanish. Since

$$\frac{\pi^2}{4}(\alpha^2 - \lambda)\prod_{n\neq 0}\left(1 + \frac{\alpha - \sqrt{\lambda}}{2n}\right)\left(1 + \frac{\alpha + \sqrt{\lambda}}{2n}\right)$$
$$= \sin\frac{\pi}{2}(\alpha - \sqrt{\lambda}) \sin\frac{\pi}{2}(\alpha + \sqrt{\lambda}),$$

it follows that the existence of a solution of type (2.52) implies the relation

$$\sin\frac{\pi}{2}(\alpha - \sqrt{\lambda})\sin\frac{\pi}{2}(\alpha + \sqrt{\lambda})D(\alpha, \lambda) = 0. \tag{2.61}$$

Therefore, if we now relate α and λ in such a manner that a solution of type (2.52) exists and, consequently, (2.61) holds, we find from (2.59) and (2.60) by a simple calculation,

$$4\sin^2\frac{\pi}{2}\sqrt{\lambda}\,D(0, \lambda) = 2 - 2\cos\pi\alpha = 2 - y_1(\pi, \lambda) - y_2'(\pi, \lambda). \tag{2.62}$$

Alternatively, we could have computed K in terms of $D(1, \lambda)$. The same argument as above would have given us the relation

$$4\cos^2\frac{\pi}{2}\sqrt{\lambda}\,D(1, \lambda) = 2 + y_1(\pi, \lambda) + y_2'(\pi, \lambda). \tag{2.63}$$

Summarizing, we have

Theorem 2.9. *The discriminant $\Delta(\lambda)$ of Hill's equation* (2.1) *can be expressed in two ways as an infinite determinant involving the Fourier coefficients g_n of $Q(x)$, (which are normalized so that $g_0 = 0$ and $g_{-n} = \bar{g}_n$); namely, with*

$$D_0(\lambda) = \left\| \frac{g_{n-m}}{\lambda - 4n^2} + \delta_{n,m} \right\|_{-\infty}^{\infty}$$

and

$$D_1(\lambda) = \left\| \frac{g_{n-m}}{\lambda - (2n+1)^2} + \delta_{n,m} \right\|_{-\infty}^{\infty},$$

we have:

$$2 - \Delta(\lambda) = 4\sin^2\left(\frac{\pi}{2}\sqrt{\lambda}\right)D_0(\lambda)$$

$$2 + \Delta(\lambda) = 4\cos^2\left(\frac{\pi}{2}\sqrt{\lambda}\right)D_1(\lambda).$$

In the case where $Q(x) = Q(-x)$, the determinants D_0 and D_1 can be factored into the product of two infinite determinants, each of which can be expressed in terms of the factors $y_1(\pi/2), \ldots, y_2'(\pi/2)$ of $\Delta - 2$ and $\Delta + 2$ [see Section 1.3, Theorem 1.2, and equations (1.10) to (1.13)]. We have

Theorem 2.10. *Let $\varepsilon_n = 2$ for $n = 1, 2, 3, \ldots$ and $\varepsilon_0 = 1$. Let*

sgn $n = +1$ *for* $n = 1, 2, 3, \ldots$, sgn $(-n) = -$sgn n *and* sgn $0 = 0$. *Then the four infinite determinants*

$$C_0(\lambda) = \left\| \delta_{n,m} + \frac{(g_{n-m} + g_{n+m})(1 + \operatorname{sgn} n \operatorname{sgn} m)}{\sqrt{\varepsilon_n \varepsilon_m}(\lambda - 4n^2)} \right\|_0^\infty$$

$$S_0(\lambda) = \left\| \delta_{n,m} + \frac{g_{n-m} - g_{n+m}}{\lambda - 4n^2} \right\|_1^\infty$$

$$C_1(\lambda) = \left\| \delta_{n,m} + \frac{(g_{n-m} + g_{n+m+1})}{\lambda - (2n+1)^2} \right\|_0^\infty$$

$$S_1(\lambda) = \left\| \delta_{n,m} + \frac{(g_{n-m} - g_{n+m+1})}{\lambda - (2n+1)^2} \right\|_0^\infty$$

(*where the* g_n *are real and* $g_n = g_{-n}$; $g_0 = 0$) *satisfy the relations*

$$C_0(\lambda) S_0(\lambda) = D_0(\lambda),$$

$$C_1(\lambda) S_1(\lambda) = D_1(\lambda),$$

$$\sqrt{\lambda} \sin\left(\frac{\pi}{2}\sqrt{\lambda}\right) C_0(\lambda) = -y_1'\left(\frac{\pi}{2}, \lambda\right),$$

$$\frac{\sin \frac{\pi}{2}\sqrt{\lambda}}{\sqrt{\lambda}} S_0(\lambda) = y_2\left(\frac{\pi}{2}, \lambda\right),$$

$$\cos\left(\frac{\pi}{2}\sqrt{\lambda}\right) C_1(\lambda) = y_1\left(\frac{\pi}{2}, \lambda\right),$$

$$\cos\left(\frac{\pi}{2}\sqrt{\lambda}\right) S_1(\lambda) = y_2'\left(\frac{\pi}{2}, \lambda\right).$$

For a proof of Theorem 2.10 see Magnus (1955).

Theorems 2.9 and 2.10 are useful for the computation of the first characteristic values of Hill's equation. Theorem 2.9 could also be used for the computation of the first terms $\Delta_n(\lambda)$ [see (2.45)] in the expansion of $\Delta(\lambda)$. However, the higher terms will then appear in a different form. For example, we find from Theorem 2.9 that

$$(2.64) \qquad \Delta_2(\lambda) = 4 \sin^2 \frac{\pi}{2}\sqrt{\lambda} \sum_{n=m+1}^{\infty} \sum_{m=-\infty}^{\infty} \frac{g_{n-m} g_{m-n}}{(\lambda - 4n^2)(\lambda - 4m^2)}$$

and it requires some nontrivial calculations to derive from this the result in Corollary 2.6.

A more detailed discussion of $\Delta(\lambda)$ and its homogeneous components

$\Delta_n(\lambda)$ may be found in Chapter VI. A different approach to the investigation of $\Delta(\lambda)$ has been developed by Hochstadt (1963a).

The analytic character of $\Delta(\lambda)$ permits the application of the Paley-Wiener Theorem and the expansion of $\Delta(\lambda)$ in a cardinal series. The results have been developed by Jagerman (1962); we list some of them here as

Jagerman's formulas. Let $\omega^2 = \lambda$. The following series converge uniformly in every closed part of the complex ω plane:

$$\begin{aligned}\Delta(\omega^2) &= 2\cos\pi\omega + \sum_{n=-\infty}^{\infty} [\Delta(n^2) - 2(-1)^n] \frac{\sin\pi(\omega - n)}{\pi(\omega - n)} \\ &= 2\cos\pi\omega + \sum_{n=-\infty}^{\infty} \left[\frac{\sin(\omega - 2n)/2}{\pi(\omega - 2n)/2}\right]^2 F_n \\ &= \Delta(0) + \omega^2 \sum_{n=-\infty}^{\infty} \frac{\Delta(n^2) - \Delta(0)}{n^2} \cdot \frac{\sin\pi(\omega - n)}{\pi(\omega - n)} \\ &= \Delta(0) + \omega^2 \sum_{n=-\infty}^{\infty} \left[\frac{\sin(\pi(\omega - 2n)/2}{\pi(\omega - 2n)/2}\right]^2 G_n,\end{aligned}$$

where

$$\begin{aligned}F_n &= \Delta(4n^2) - 2 + (\omega - 2n)2n\Delta'(4n^2), \\ G_n &= \left(\frac{\omega}{n} - 1\right)\frac{\Delta(4n^2) - \Delta(0)}{4n^2} + \left(\frac{\omega}{n} - 2\right)\Delta'(4n^2),\end{aligned}$$

and Δ' denotes the derivative of Δ with respect to ω.

2.4. *Asymptotic behavior of the characteristic values*

In this section we shall be concerned with estimates (upper and lower bounds) for the characteristic values λ_n ($n = 0, 1, \ldots$) and λ_m' ($m = 1, 2, 3, \ldots$) or, alternatively, with the location of the intervals of stability as defined in Theorem 2.1 of Section 2.1.

We shall not prove the sharpest results available. For a detailed account of the finer points of the theory, we refer to Chapter V and to the monographs by Starzinskiĭ (1955) and Krein (1955) which also contain an extensive bibliography, and also to the paper by Borg (1946). What we shall prove here are results first found by Borg and proved again in a much simpler manner by Hochstadt (1961, 1963b).

We begin by proving a preliminary result which has been tacitly assumed to be true in most papers on the subject but which seems not to have been stated explicitly anywhere. We shall derive

Lemma 2.9. Let $Q(x)$ in (2.1) be a bounded integrable function of x with period π satisfying (2.41), and let λ_n $(n = 0, 1, 2, \ldots)$ and λ_m' $(m = 1, 2, 3, \ldots)$ be defined as in Theorem 2.1. Then there exist sequences of nonnegative numbers ε_n and ε_m' such that the sequences

$$\sqrt{n}\,\varepsilon_n, \quad \sqrt{m}\,\varepsilon_m' \tag{2.65}$$

are bounded and

$$|\sqrt{\lambda_{2n-1}} - 2n| < \varepsilon_n, \quad |\sqrt{\lambda_{2n}} - 2n| < \varepsilon_n, \tag{2.66}$$

$$|\sqrt{\lambda'_{2m-1}} - (2m - 1)| < \varepsilon_m', \quad |\sqrt{\lambda_{2m}'} - (2m - 1)| < \varepsilon_m'. \tag{2.67}$$

Proof. It is not difficult to show that, for large λ, the square roots of the zeros λ_n of $\Delta(\lambda) - 2$ lie in the neighborhood of the even integers and that the square roots of the zeros λ_m' of $\Delta(\lambda) + 2$ lie in the neighborhood of the odd integers. In fact, we have from Theorem 2.4:

$$|\Delta(\lambda) - 2 + 4\sin^2(\pi\sqrt{\lambda}/2)| \leq B/\sqrt{\lambda}, \tag{2.68}$$

where B is independent of λ. If $\Delta(\lambda) = 2$, and if λ is large, then $\sin(\pi\sqrt{\lambda}/2)$ must be small. Putting

$$\sqrt{\lambda} = 2n + \eta_n, \qquad -1 \leq \eta_n \leq 1,$$

where n is a positive integer, and using the fact that

$$\sin^2(t\pi/2) \geq t^2 \quad \text{for} \quad -1 \leq t \leq 1$$

we find

$$\eta_n^2 \leq B/(2n - 1)$$

and therefore, with a suitable B^* and for $n \geq 1$:

$$|\eta_n| \leq B^*/\sqrt{n} \qquad (n = 1, 2, 3, \ldots)$$

A similar argument holds for the roots of $\Delta(\lambda) + 2 = 0$. However, this does not prove Lemma 2.9 completely. What is missing is a proof of the fact that the two roots of $\Delta(\lambda) - 2 = 0$, the square roots of which are, for large n, arbitrarily close to $2n$ (or, correspondingly, arbitrarily close to $2m - 1$ for large m) are *exactly* those denoted by λ_{2n-1} and

λ_{2n} (or, in the case of $\Delta(\lambda) + 2 = 0$, by λ'_{2m-1} and λ'_{2m}). To prove these facts, thereby completing the proof of Lemma 2.9, we shall show:

(i). Let $2N + 1$ be an odd positive integer. Then, if N is sufficiently large, the number of zeros of the analytic functions

$$\Delta(\lambda) - 2, \quad \sin^2(\pi\sqrt{\lambda}/2)$$

is the same in the half-plane in which the real part of λ is less than $(2N + 1)^2$.

(ii). Let $2N + 2$ be an even positive integer. Then, if N is sufficiently large, the analytic functions

$$\Delta(\lambda) + 2, \quad \cos^2(\pi\sqrt{\lambda}/2)$$

have the same numbers of zeros in that part of the λ plane in which the real part of λ is less than $4(N + 1)^2$.

In both statements (i) and (ii), multiple zeros are to be counted properly.

We shall prove statement (i) only, since the proof of (ii) would not involve any new arguments.

Let $\omega = \sqrt{\lambda}$ and consider the path P in the right half of the ω plane consisting of three sides A, B, and C of a square which, respectively, are defined by

$$\begin{aligned} &A\colon \omega = t - i(2N + 1), && 0 \le t \le 2N + 1 \\ &B\colon \omega = 2N + 1 + is, && -(2N + 1) \le s \le 2N + 1 \\ &C\colon \omega = \tau + i(2N + 1), && 2N + 1 \ge \tau \ge 0. \end{aligned}$$

Under the mapping defined by $\lambda = \omega^2$, P will be mapped onto a closed, piecewise differentiable curve C^* in the λ plane which contains in its interior the segment S

$$-(2N + 1)^2 \le \lambda \le (2N + 1)^2$$

of the real λ axis. Let $\theta(\lambda)$ denote the meromorphic function of λ defined by

$$\theta(\lambda) = (-\Delta(\lambda) + 2)/[4 \sin^2(\pi\sqrt{\lambda}/2)],$$

and let Z be the integral

$$Z = \frac{1}{2\pi i}\int_{C^*} \frac{d}{d\lambda} \log \theta(\lambda)\, d\lambda.$$

Then Z is the difference between the number of zeros of $\Delta(\lambda) - 2$ and of $\sin^2(\pi\sqrt{\lambda}/2)$ in the interior of C^* or (since all of these zeros are real), the difference of these numbers on the segment S. Statement (i) is equivalent to saying that $Z = 0$. (Of necessity, Z is an integer.)

Using (2.38) in Theorem 2.4, we find that on the whole path

$$|\theta - 1| \leq \frac{\beta_N}{1 - 2\alpha_N};$$

where

$$\beta_N = 2|\exp(\pi M^*/(2N + 1)) - 1|, \quad \alpha_N = e^{-\pi(2N+1)}.$$

We conclude that, for sufficiently large N, the map of the curve C^* in the θ plane is entirely within a circle of radius <1, with center at $\theta = 1$. Therefore, the integral Z can be evaluated by taking the difference of the values of the integrand at the end points of C^* (which coincide in the λ plane but at which the values of $\log\theta$ may be different). Again, for sufficiently large N, these values of $\log\theta$ must be less than ½ in absolute value and, since Z is an integer, Z must be zero. This proves Lemma 2.9.

A result stronger than Lemma 2.9 has been proved by Borg (1944, 1946) under similar assumptions about $Q(x)$. We state it without proof as

***Theorem 2.11* (G. Borg).** *If $Q(x)$ in* (2.1) *is periodic with period π such that*

$$\frac{1}{\pi}\int_0^\pi |Q(x)|\,dx = A \tag{2.69}$$

exists, and such that the normalizing condition (2.41) *is satisfied, then, for any integer $n > A/2$,*

$$|\sqrt{\lambda_{2n-1}} - 2n| \leq A/(4n), \quad |\sqrt{\lambda_{2n}} - 2n| \leq A/(4n), \tag{2.70}$$

$$\begin{aligned} &|\sqrt{\lambda'_{2n-1}} - 2n + 1| \leq A/(4n - 2), \\ &|\sqrt{\lambda'_{2n}} - 2n + 1| \leq A/(4n - 2). \end{aligned} \tag{2.71}$$

Theorem 2.11 has been supplemented by Borg by an additional theorem which sharpens the inequalities (2.70) and (2.71) according to the nature of $Q(x)$. His results are, in a certain sense, best possible. We shall not consider this latter aspect of Borg's theory. Instead, we shall derive some of his improvements of Theorem 2.11 which are valid for sufficiently differentiable $Q(x)$, using a very simple method discovered by Hochstadt (1963b). We have:

Theorem 2.12. *Let $Q(x)$ in (2.1) be periodic with period π and assume that the second derivative Q'' exists and is continuous. Let $Q(x)$ be normalized according to (2.41), i.e., such that*

$$\int_0^\pi Q(x)\,dx = 0,$$

and let C be defined by

$$C = \frac{1}{\pi}\int_0^\pi Q^2(x)\,dx. \tag{2.72}$$

Then for large n,

$$\begin{aligned} \lambda'_{2n-1} - (2n-1)^2 - C/(4n)^2 &= o(n^{-2}) \\ \lambda'_{2n} - (2n-1)^2 - C/(4n)^2 &= o(n^{-2}) \\ \lambda_{2n-1} - 4n^2 - C/(4n)^2 &= o(n^{-2}) \\ \lambda_{2n} - 4n^2 - C/(4n)^2 &= o(n^{-2}). \end{aligned} \tag{2.73}$$

Proof. We shall consider only such values of λ for which

$$\lambda + Q(x) > 0, \quad [\lambda + Q(x)]^{3/2} > \tfrac{1}{4}|Q'(x)|, \tag{2.74}$$

for all x. Then we may write the general solution $y(x)$ of (2.1) and its derivative $y'(x)$ in the form

$$y(x) = A(x)\sin\varphi(x), \quad y'(x) = [\lambda + Q(x)]^{1/2}A(x)\cos\varphi(x) \tag{2.75}$$

where the functions $\varphi(x)$ and $A(x)$ are determined by their initial values $\varphi(0)$, $A(0)$, and the differential equations

$$\varphi'(x) = [\lambda + Q(x)]^{1/2} + \frac{Q'(x)\sin 2\varphi(x)}{4[\lambda + Q(x)]} \tag{2.76}$$

and

$$\frac{d}{dx}\log A(x) = -\frac{Q'(x)\cos^2\varphi(x)}{2[\lambda + Q(x)]} \tag{2.77}$$

In fact, it is easy to see that (2.76) and (2.77) together with (2.1) imply (2.75) and vice versa. Also, the first inequality in (2.74) guarantees that the solutions $\varphi(x)$ of (2.76) are bounded for all finite values of x, and the second inequality in (2.74) guarantees that $\varphi(x)$ is a monotonically increasing function of x. Furthermore, (2.77) shows that $A(x)/A(0)$ is a positive function of x, and finally, we can use $\varphi(0)$ and $A(0)$ to give prescribed values to $y(0)$ and $y'(0)$.

Suppose now that $y(x)$ is periodic with period π or 2π. In the first case, we have from (2.75) that

$$A(\pi) = A(0), \quad \varphi(\pi) - \varphi(0) = 2n\pi \tag{2.78}$$

where n is a positive integer. In the second case, we use Flocquet's theorem (Section 1.2) which shows that $y(\pi) = -y(0)$ and $y'(\pi) = -y'(0)$, and derive from (2.75) that

$$A(\pi) = A(0), \quad \varphi(\pi) - \varphi(0) = (2n - 1)\pi, \tag{2.79}$$

where n is again a positive integer.

From now on we shall confine ourselves to the case where $y(x + \pi) = y(x)$; the case of a solution of (2.1) of period 2π can be treated in exactly the same manner. From (2.76) and (2.79) we obtain for the values of λ belonging to a solution of period π

$$2n\pi = \int_0^\pi [\lambda + Q(x)]^{1/2}\, dx + \int_0^\pi \frac{Q'(x) \sin 2\varphi(x)}{4[\lambda + Q(x)]}\, dx. \tag{2.80}$$

We know from Lemma 2.9 that, for sufficiently large n, (2.80) must have two solutions λ_{2n-1} and λ_{2n} which are "close" to $(2n)^2$ in the sense defined by Lemma 2.9. In order to utilize our assumption about $Q(x)$, we integrate (2.80) by parts and obtain

$$\begin{aligned}
2n\pi &= \sqrt{\lambda} \int_0^\pi [1 + Q(x)/\lambda]^{1/2}\, dx - \frac{1}{8\lambda} \int_0^\pi \frac{Q'(x)(\cos 2\varphi(x))'}{[1 + Q(x)/\lambda]\varphi'(x)}\, dx \\
&= \pi\sqrt{\lambda} + \tfrac{1}{2}\lambda^{-1/2} \int_0^\pi Q(x)\, dx \\
&\quad - \tfrac{1}{8}\lambda^{-3/2} \int_0^\pi Q^2(x)\, dx + R_1\lambda^{-2} \\
&\quad - \left|\frac{1}{8\lambda} \frac{Q'(x) \cos 2\varphi(x)}{[1 + Q(x)/\lambda]\varphi'(x)}\right|_0^\pi \\
&\quad + \frac{1}{8\lambda^{3/2}} \int_0^\pi Q''(x) \cos 2\varphi(x)\, dx + R_2\lambda^{-2},
\end{aligned} \tag{2.81}$$

where $|R_1|$ and $|R_2|$ are bounded independently of λ for all sufficiently large λ and are given by

$$\lambda^{-2}R_1 = \sqrt{\lambda}\int_0^\pi \{[1 + Q(x)/\lambda]^{1/2} - 1 - \tfrac{1}{2}Q(x)/\lambda + \tfrac{1}{8}Q^2(x)/\lambda^2\}\,dx$$

$$\lambda^{-2}R_2 = \frac{1}{8\lambda}\int_0^\pi \cos 2\varphi(x)\left\{\frac{d}{dx}\frac{Q'(x)}{\varphi'(x)[1 + Q(x)/\lambda]} - \frac{Q''(x)}{\sqrt{\lambda}}\right\}dx.$$

The boundedness of R_1 is recognized immediately since R_1 is the integral of the remainder term of a convergent Taylor series. In order to recognize the boundedness of R_2, we have to observe that, for $\lambda \to \infty$, $\varphi'(x)$ behaves like $\sqrt{\lambda}$ and $\varphi''(x)$ behaves like $1/\sqrt{\lambda}$. Returning now to (2.81) and using (2.41) and the periodicity of the continuous function $Q'(x)$, we obtain

$$(2.82)\quad 2n\pi = \pi\sqrt{\lambda} - \frac{\pi}{8}C\lambda^{-3/2} + \frac{\lambda^{-3/2}}{8}\int_0^\pi Q''(x)\cos 2\varphi(x)\,dx + \mathcal{O}(\lambda^{-2})$$

where C is defined by (2.72). Our next task will be to prove that, for $\lambda \to +\infty$

$$(2.83)\qquad \int_0^\pi Q''(x)\cos 2\varphi(x)\,dx = o(1).$$

For this purpose, we introduce the functions

$$\rho(x) = \varphi(x) - \varphi(0) - x\sqrt{\lambda} = \varphi(x) - \beta(x),\quad \beta(x) = x\sqrt{\lambda} + \varphi(0),$$

and observe that (2.76) gives us the relation

$$(2.84)\qquad |\rho(x)| \le \theta\lambda^{-1/2}\qquad (\theta = \text{constant}).$$

because $\sin 2\varphi$ is uniformly bounded for all real $\varphi(x)$. Now we have

$$\int_0^\pi Q''(x)\cos 2\varphi(x)\,dx = \int_0^\pi Q''(x)\cos 2[\sqrt{\lambda}x + \varphi(0)]\,dx$$
$$+ \int_0^\pi Q''(x)[-2\sin^2\rho(x)\cos 2\beta(x)$$
$$+ \sin 2\rho\sin 2\beta]\,dx$$

In the last equation, the first integral goes to zero as $\lambda \to \infty$ because of the Riemann-Lebesgue theorem, and the second integral tends towards zero because of (2.84). Therefore, we find from (2.82) that λ_{2n} and λ_{2n-1} must satisfy the relation for λ:

$$2n\pi = \pi\lambda^{1/2} - (\pi/8)C\lambda^{-3/2} + o(\lambda^{-3/2}).$$

Using now Lemma 2.9 we find, with $\sqrt{\lambda} = 2n + \varepsilon$,

$$0 = \varepsilon - \tfrac{1}{8}\, C(2n)^{-3} + o(n^{-3})$$

which gives us immediately Theorem 2.12.

An application of Theorem 2.12 may be found in Chapter III, after Lemma 3.1.

As a consequence of Theorem 2.12, we have an estimate for the differences

$$|\lambda_{2n} - \lambda_{2n-1}|, \quad |\lambda'_{2n} - \lambda'_{2n-1}|$$

for large n. It is clear that we can improve on the estimates if we can carry out more than one integration by parts (of the type used in the proof of Theorem 2.12). For details see Hochstadt (1963b). We state some of the results as

Theorem 2.13. *If Q'' in* (2.1) *is continuous, the intervals of instability d_n and d_n', defined by*

$$d_n = |\lambda_{2n} - \lambda_{2n-1}|, \quad d_n' = |\lambda'_{2n} - \lambda'_{2n-1}|$$

tend to zero for $n \to \infty$ in such a manner that

$$\lim_{n\to\infty} n^2\, d_n = 0, \quad \lim_{n\to\infty} n^2\, d_n' = 0.$$

If $Q(x)$ is analytic for real x, then, for all exponents κ = 2, 3, 4, . . .

$$\lim_{n\to\infty} n^\kappa\, d_n = 0, \quad \lim_{n\to\infty} n^\kappa\, d_n' = 0.$$

2.5. *Basic results from the general theory of linear differential equations*

In this section, we review a few results of a fundamental nature which do not belong specifically to the theory of Hill's equation but are relevant to it. Terms belonging to the general theory of linear differential equations will be used without further explanation.

The following result has been proved by Haupt (1914)

Theorem 2.14. *Let $y(x, \lambda)$ be a nontrivial, real periodic solution of* (2.1) *with period π or 2π. If $\lambda = \lambda'_{2n+1}$ or $\lambda = \lambda'_{2n}$, then y has exactly*

$2n + 1$ zeros in the half-open interval $0 \leq x < 2\pi$. If $\lambda = \lambda_{2n-1}$ or $\lambda = \lambda_{2n}$, then y has exactly $2n$ zeros in $0 \leq x < \pi$.

The question of for which values of λ any solution of (2.1) will have infinitely many zeros is discussed in Chapter IV. See also Yelchin (1946) and Starzinskiĭ (1955).

Periodic solutions of period π belonging to different characteristic values λ_n and λ_m of (2.1) are orthogonal in the interval $(0, \pi)$. Any two periodic solutions of period π or 2π are orthogonal in the interval $(0, 2\pi)$ if they belong to different characteristic values.

The following result is the analog of the well-known theorem about the expansion of a function in a Fourier series. [See Weyl (1910) or Coddington and Levinson (1955).] We have

Theorem 2.15. *Let μ_m, $m = 0, 1, 2, \ldots$ be the characteristic values $\lambda_0, \lambda_1', \lambda_2', \lambda_1, \ldots$, in their natural order and let $z_m(x)$ be an ortho-normal set of periodic solutions of (2.1) such that z_m satisfies the equation*

$$z_m'' + [\mu_m + Q(x)]z = 0.$$

Then every continuous periodic function of period 2π whose second derivative is square integrable in every finite interval can be expanded in a uniformly and absolutely convergent series

$$\sum_{m=0}^{\infty} c_m z_m(x)$$

with constant coefficients c_m.

The theory of Weyl (1910) can also be applied to the differential equation (2.1) for the interval $-\infty < x < \infty$. There always exists a solution of (2.1) which is not square integrable in $(-\infty, 0)$. An analogous statement is valid for the interval $(0, \infty)$. We thus have the limit point case at both end points $-\infty$ and $+\infty$. Since every non-trivial solution of (2.1) fails to be square integrable in $(-\infty, \infty)$ (cf. Floquet's theorem), the spectrum is purely continuous and can be seen to coincide with the union of the intervals of stability [see Hartman and Wintner (1949)]. Thus, according to Weyl (1910), we have the following

Theorem 2.16. *Let $f(x)$ be a continuous and twice-differentiable function of x which is defined for $-\infty < x < \infty$ and for which*

$$\int_{-\infty}^{\infty} |f(x)|^2 \, dx < \infty, \quad \int_{-\infty}^{+\infty} |f''(x)|^2 \, dx < \infty.$$

Let S be the set on the real λ axis *which consists of the union of the open intervals of stability of* (2.1) *and their end points. Let $y_1(x, \lambda)$ and $y_2(x, \lambda)$ be the normalized solutions of* (2.1). *Then there exist functions $M_1(\lambda)$ and $M_2(\lambda)$ defined on S and such that*

$$f(x) = \int_S \{y_1(x, \lambda)\, dM_1(\lambda) + y_2(x, \lambda)\, dM_2(\lambda)\}.$$

According to Borg (1946), the only case in which the set S is connected occurs if $Q(x) \equiv 0$. This case leads to the ordinary Fourier theorem.

The question for which $Q(x)$ the set S will consist of exactly two pieces has been settled by Hochstadt (1965).

For a full discussion of these two cases see Section 7.6.

Schaefke (1960/61) considered the case where $Q(x)$ is a holomorphic function of the complex variable x in a strip containing the real x axis and bounded by two straight lines parallel to it. He derives series expansion for functions $f(x)$ holomorphic in the same strip and satisfying the functional equation

$$f(x + \pi) = \exp(\pi i\nu) f(x), \qquad (\nu \text{ arbitrary complex})$$

in terms of those solutions of (2.1) which satisfy the same functional equation. These belong to values of λ which are roots of the transcendental equation

$$\Delta(\lambda) = 2 \cos \nu\pi.$$

The results obtained by Schaefke are particularly useful in deriving expansion theorems involving the classical functions of mathematical physics (Bessel, Whittaker, Legendre functions, and others).

The last theorem to be mentioned here concerns the determination of the function $Q(x)$ by the spectrum, i.e., by the values μ_m of Theorem 2.15. Strictly speaking, it is a theorem on Hill's equation only if $Q(x)$ is an even function of x, in which case the conditions for the existence of periodic solutions can be formulated as boundary conditions (see Theorem 1.2, Section 1.3). The following result is due to Borg (1946). A relatively simple proof may be found in Levin (1964) (Appendix 4.2, pp. 430–436). We have

Theorem 2.17. *Let $i = 1, 2$, let h and h_i be real constants and let Q be a real-valued function for which the Lebesgue integral*

$$\int_0^\pi |Q(x)|\, dx < \infty.$$

Let S_i (h, h_i) be the set of values of λ for which the differential equation

$$y'' + [\lambda + Q(x)]y = 0$$

has a nontrivial solution satisfying

$$y'(0) - hy(0) = 0, \quad y'(1) + h_i y(1) = 0.$$

Then the sets $S_1(h, h_1)$ and $S_2(h, h_2)$ determine $Q(x)$ for $0 \leq x \leq \pi$ almost everywhere and, in addition, determine h, h_1, and h_2 uniquely.

2.6. Theorems of Liapounoff and Borg and the Fourier transform

In this section, we shall state without proof a few results concerning Hill's equation which are of interest for different reasons. Liapounoff's theorem, which we shall state first, is the earliest example of a type of result which guarantees stability of the solutions of Hill's equation without involving data which can be derived only by actually solving the equation. Liapounoff's result has been generalized in many directions; some of the later results will be presented in Sections 5.2; others may be found in the monographs by Starzimskiĭ (1954), and Krein (1951). We have, according to Liapounoff (1907)

Liapounoff's Theorem. *Let $p(x) \neq 0$ be a nonnegative, piecewise-continuous, periodic function with period π. Then all solutions of*

$$y'' + p(x)y = 0$$

are bounded for all values of x if

$$\pi \int_0^\pi p(x)\, dx \leq 4.$$

This condition is best possible in the sense that, for any $\varepsilon > 0$, there exists a nonnegative, piecewise-continuous function $p_0(x)$ satisfying

$$p_0 \not\equiv 0, \quad p_0(x + \pi) = p_0(x)$$

and

$$\pi \int_0^\pi p_0(x)\, dx < 4 + \varepsilon,$$

such that at least one solution of

$$y'' + p_0(x)y = 0$$

is unbounded as $x \to \pm\infty$.

Generalizations of Theorem 2.13, which characterize the nth interval of stability of (2.1) and are best possible results in the same sense as Theorem 2.13, are due to Borg (1944) (see Section 5.2).

The next theorem is one of the many results in the paper by Borg (1946). We state it here not only because of its many applications but also because its proof has withstood so far all attempts at simplification. It belongs in the context of the "coexistence problem", i.e., of the problem to determine when two linearly independent periodic solutions of (2.1) of period π or 2π can coexist. We have

Borg's theorem. *All the roots of the equation $\Delta(\lambda) + 2 = 0$ are double roots (and two linearly independent solutions of (2.1) of period 2π exist whenever one such solution exists) if and only if $Q(x)$ has period $\pi/2$.*

The "if" part of this result is contained in the trivial corollary to Floquet's theorem in Section 1.2. An immediate consequence of the theorem is the result that, if the roots of both $\Delta(\lambda) + 2$ and of $\Delta(\lambda) - 2$, with the exception of λ_0, are double roots, then $Q(x)$ must be a constant since it must have periods $\pi/2$, $\pi/4$, $\pi/8$, etc. This consequence can now be proved quite simply [Ungar (1961)] and the details may be found in Section 7.6.

Finally, we mention briefly that the theory of the Fourier transformation can be applied to Hill's equation.

The method used in proving Theorem 2.2 can also be used to show that for every fixed value of x a nontrivial solution $y(x, \lambda)$ of (2.1) is of order of growth $\frac{1}{2}$ with respect to λ. From the estimates (2.32) and from the theorem due to Paley and Wiener (1934) (Theorem 10, p. 13) the following fact can be derived: Let x be real and let $y(x, \lambda)$ be a solution of (2.1) for which

$$y(0, \lambda) = a, \quad y'(0, \lambda) = b.$$

Then there exists a function $G(x, \theta)$ of the real variables x and θ which is defined for $|\theta| \leq |x|$ and is such that for all real values of x and for $\lambda > 0$

$$y(x, \lambda) = a \cos(x\sqrt{\lambda}) + \int_{-x}^{x} G(x, \theta)e^{i\theta\sqrt{\lambda}}\, d\theta.$$

The function $G(x, \theta)$ satisfies the partial differential equation

$$\frac{\partial^2 G}{\partial x^2} - \frac{\partial^2 G}{\partial \theta^2} + Q(x)G = 0$$

For details see Magnus (1955); for applications, see Gelfand and Levitan (1955) (p. 296) and Levin (1964) (Appendix 4.1, pp. 426–429).

For the discriminant of Hill's equation, Jagerman (1962) derived from the Paley-Wiener theorem formulas which have been stated at the end of Section 2.3.

PART II

DETAILS

It is the purpose of the second part of this monograph to provide some aids for anyone who wishes to apply Hill's equation and also to discuss some finer points of the general theory. In particular, we shall present a few ready-made transformation and approximation formulas, and we shall discuss some examples in detail. Also, we shall discuss whatever is known about the problem of coexistence of solutions of period π or 2π (i.e., the collapsing of intervals of instability) and we shall state a fair number of inequalities, partly of the type of Liapounoff's theorem and partly of the type of estimates as given in Theorem 2.12.

III

Elementary Formulas

3.1. *Transformation into a standard form*

The differential equation

$$\frac{d^2z}{dx^2} + a(x)\frac{dz}{dx} + b(x)z = 0 \tag{3.1}$$

can be transformed into

$$\frac{d^2y}{dx^2} + Q(x)y = 0 \tag{3.2}$$

but putting

$$y = [\exp \tfrac{1}{2}A(x)]z, \quad a(x) = dA/dx, \tag{3.3}$$

where

$$Q = -\frac{1}{2}\frac{da}{dx} - \frac{1}{4}a^2 + b. \tag{3.4}$$

3.2. *The Liouville transformation*

The differential equation for the interval $0 \leq t \leq \omega$:

$$\frac{d^2z}{dt^2} + \lambda M^4(t)z = 0 \qquad (\lambda = \text{constant}, M(t) > 0) \tag{3.5}$$

can be transformed into the differential equation

$$\frac{d^2y}{dx^2} + [\lambda\gamma^2 + Q(x)]y = 0 \tag{3.6}$$

for the interval $0 \leq x \leq \pi$ by the *Liouville transformation:*

(3.7) $$x = \frac{1}{\gamma}\int_0^t M^2(\tau)\,d\tau, \quad \gamma = \frac{1}{\pi}\int_0^\omega M^2(t)\,dt,$$

(3.8) $$y(x) = M(t)z(t),$$

(3.9) $$Q(x) = [M(t)]^{-1}\frac{d^2M(t)}{dx^2} = -\gamma^2[M(t)]^{-3}\frac{d^2}{dt^2}\left(\frac{1}{M(t)}\right).$$

The inverse of this transformation is given by

(3.10) $$t = \gamma\int_0^x \frac{d\zeta}{[M^*(\zeta)]^2}, \quad \gamma^{-1} = \omega^{-1}\int_0^\pi \frac{d\xi}{[M^*(\xi)]^2},$$

where $M^*(x) = M(t)$ is a positive solution of the differential equation

(3.11) $$\frac{d^2M^*}{dx^2} = Q(x)M^*(x).$$

It should be noted that the Liouville transformation is not applicable unless $M(t)$ is twice differentiable, at least in the sense that d^2M/dt^2 is bounded and continuous almost everywhere. An example of the Liouville transformation is provided by the following

Lemma 3.1. The function $Q(x)$ corresponding to $M(t)$ under the Liouville transformation is a constant if and only if

(3.12) $$M(t) = [\alpha t^2 + 2\beta t + \delta]^{-1/2}, \qquad (\alpha, \beta, \delta \text{ constant}).$$

In this case the differential equation for $y(x)$ will be

(3.13) $$\frac{d^2y}{dx^2} + \gamma^2[\lambda - (\alpha\delta - \beta^2)]y = 0.$$

If

(3.14) $$D^2 = \alpha\delta - \beta^2 > 0,$$

that is, if $\alpha t^2 + 2\beta t + \delta$ does not have any real zeros, we have

(3.15) $$\pi D\gamma = \tan^{-1}[(\alpha\omega + \beta)D^{-1}] - \tan^{-1}(\beta D^{-1}).$$

Obviously,

$$|D\gamma| \leq 1.$$

As an application of Lemma 3.1, we shall answer a question raised by Mothwurf in a manner proposed by Borg (1946). If we consider an inhomogeneous vibrating string with the mass distribution $M^4(t)$, we may ask for which functions $M(t)$ will all overtones be exact integral multiples of the lowest tone? If we consider the case where the end points of the string are kept in a fixed position, the mathematical equivalent of this problem is the question: For which $M(t)$ are the square roots of the eigenvalues $\mu_n = \sqrt{\lambda_n^*}\,(n = 1, 2, 3, \ldots)$ of the boundary-value problem $z(0) = z(\omega) = 0$ of equation (3.5) integral multiples of $\sqrt{\lambda_1^*}$?

The answer given by Borg is the following one: Even if we merely postulate that infinitely many of the μ_n are integral multiples of one of them, $M(t)$ must be of the form (3.12). In this case, it is automatically true that *all* μ_n are integral multiples of μ_1.

To prove this statement, we transform (3.5) into (3.6). We then have the problem to determine $Q(x)$ in such a manner that infinitely many of the roots λ_n^* of the equation $y_2(\pi, \lambda) = 0$ [where $y_2(0, \lambda) = 0$; $y_2'(0, \lambda) = 1$] are integral multiples of one of these roots. Now we observe that the λ_n^* are all located in the intervals of instability of Hill's equation (3.6). For, if $y_2(\pi, \lambda^*) = 0$, we have from

$$y_1 y_2' - y_2 y_1' = 1$$

that $y_1(\pi, \lambda^*) y_2'(\pi, \lambda^*) = 1$. Therefore,

$$|\Delta(\lambda^*)| = |y_1(\pi, \lambda^*) + y_2'(\pi, \lambda^*)| \geq 2,$$

and therefore the λ_n^* are in the intervals of instability of (3.6). But according to Theorem 2.12, these intervals have end points very close to the points

$$m^2 + Cm^{-2},$$

where m is an integer and

$$C = \frac{1}{\pi}\int_0^\pi Q_0^2(x)\,dx,$$

where Q_0 differs from our Q by an additive constant. Now it is seen easily that the conditions imposed on the $\sqrt{\lambda_n^*}$ can be satisfied only if $C = 0$ and therefore, $Q_0 \equiv 0$.

3.3. *Polar coordinates*

Let $y_1(x)$ and $y_2(x)$ be the standard solutions of

$$\frac{d^2y}{dx^2} + Q(x)y = 0 \tag{3.16}$$

which satisfy the initial conditions

$$y_1(0) = 1, \quad y_2(0) = 0, \quad y_1'(0) = 0, \quad y_2'(0) = 1, \tag{3.17}$$

where a prime denotes the derivative with respect to x. If we interpret y_1 and y_2 as Cartesian coordinates in a plane, a *transformation to polar coordinates* leads to the following formulas: Let

$$y_1(x) = \rho \cos \phi, \quad y_2(x) = \rho \sin \phi, \tag{3.18}$$
$$\rho > 0, \quad \phi(0) = 0, \quad \rho(0) = 1.$$

Then

$$\phi(x) = \int_0^x \frac{dt}{\rho^2(t)}, \tag{3.19}$$

$$\rho^2(x) = y_1^2(x) + y_2^2(x), \tag{3.20}$$

$$\frac{d^2\rho}{dx^2} - \rho^{-3} + Q(x)\rho = 0. \tag{3.21}$$

It should be noted that $\phi(x)$ is a monotonically increasing function of x. If $|\phi(x)| \to \infty$ as $x \to +\infty$ or $x \to -\infty$, both $y_1(x)$ and $y_2(x)$ will have infinitely many zeros. Otherwise, both of them will have a finite number of zeros.

3.4. *Differential equation for the product of two solutions*

Let $y = \eta_1$ and $y = \eta_2$ be any two solutions of

$$\frac{d^2y}{dx^2} + Q(x)y = 0, \tag{3.22}$$

where, for all x,

$$Q(x + \pi) = Q(x).$$

Let

$$z = \eta_1\eta_2$$

be the product of these solutions. Then

$$\frac{d^3z}{dx^3} + 4Q\frac{dz}{dx} + 2\frac{dQ}{dx}z = 0, \tag{3.23}$$

and equation (3.23) has at least one nontrivial periodic solution with period π. The following result holds

Lemma 3.2. Either all periodic solutions of (3.23) with period π are constant multiples of a single one, or all solutions of (3.23) are periodic with period π. This takes place if and only if all solutions of (1.16) are periodic with period π or 2π.

Lemma 3.2 is an immediate consequence of Floquet's theorem (see Part I, Section 1.2). If the characteristic equation for (3.16) has the roots ρ_1 and ρ_2, and if η_1 and η_2 are nontrivial solutions such that

$$\eta_1(x + \pi) = \rho_1\eta_1(x), \quad \eta_2(x + \pi) = \rho_2\eta_2(x), \tag{3.24}$$

then clearly $\eta_1\eta_2$ is periodic with period π since $\rho_1\rho_2 = 1$. In general, all periodic solutions of (3.23) with period π will be multiples of this particular one because of the following rather obvious fact:

If η_1 and η_2 are two linearly independent solutions of (1.16), then

$$\eta_1^2, \quad \eta_1\eta_2, \quad \eta_2^2$$

are linearly independent solutions of (3.23). That all solutions of (3.23) will have period π if and only if all solutions of (1.16) are periodic with period π or 2π can be derived from this fact by choosing for η_1 and η_2 the linearly independent solutions which appear in Floquet's theorem.

Equation (3.23) can be reduced to the nonlinear second order equation

$$z\frac{d^2z}{dx^2} - \frac{1}{2}\left(\frac{dz}{dx}\right)^2 + 2Qz^2 = C \tag{3.25}$$

where C is a constant. Putting $z = \eta_1\eta_2$, we find

$$C = -\frac{1}{2}\left(\eta_1\frac{d\eta_2}{dx} - \eta_2\frac{d\eta_1}{dx}\right)^2,$$

and this implies

Lemma 3.3. Assume that not all of the solutions of (3.23) are periodic with period π. Then, if z is a periodic nontrivial solution, the constant C in (3.25) will be:

(1) negative, if $y'' + Qy = 0$ has unbounded solutions;
(2) positive, if all solutions are bounded;
(3) zero, if a nontrivial solution has period π or 2π.

IV

Tests for the Existence of Oscillating Solutions

The real solutions of Hill's equation with a real λ and Q:

$$(4.1) \quad \frac{d^2y}{dx^2} + [\lambda + Q(x)]y = 0, \quad \int_0^\pi Q(x)\,dx = 0, \quad Q(x + \pi) = Q(x),$$

may or may not have infinitely many zeros. The situation can be described as follows:

Theorem 4.1. *Either all nontrivial real solutions of* (4.1) *have only a finite number of zeros, or all real solutions of* (4.1) *have infinitely many zeros. Let λ_0 be the smallest value of λ for which* (4.1) *has a periodic solution. Then for $\lambda \leq \lambda_0$, all nontrivial real solutions have only a finite number of zeros, but for $\lambda > \lambda_0$, every real solution has infinitely many zeros.*

We shall prove Theorem 4.1 by using part of a result which is due to Hamel (1912).

Theorem 4.2. *There exists a real solution of* (4.1) *which has only a finite number of zeros if and only if for all continuously differentiable functions $w(x)$ with period π*

$$(4.2) \qquad \int_0^\pi [\lambda + Q(x)]w^2(x)\,dx \leq \int_0^\pi \left(\frac{dw}{dx}\right)^2 dx.$$

We shall need only the fact that (4.2) must hold if (4.1) has a solution without infinitely many zeros, and we shall not prove the sufficiency of condition (4.2) (see Lemma 4.2).

To prove Theorem 4.1, we observe first that every real solution of (4.1) will have infinitely many zeros if a single nontrivial solution has this property. This follows from the formulas which describe the transformation to polar coordinates. In fact, (3.18) shows that every real solution of (4.1) can be written in the form

$$y(x) = A\rho(x)\cos(\phi - \phi_0) \tag{4.3}$$

where A and ϕ_0 are constants. If (4.3) has infinitely many real zeros for $A \neq 0$ and a particular value of ϕ_0, then $\phi(x)$ must be unbounded if x goes from $-\infty$ to $+\infty$, and therefore $y(x)$ has infinitely many zeros for every choice of ϕ_0 and A. Now we can prove

Lemma 4.1. If (4.1) has a real, nontrivial solution with finitely many zeros, then (4.1) must have a real solution $y(x)$ such that

$$y(x + \pi) = \rho y(x), \tag{4.4}$$

where ρ is real and positive and where $y(x)$ has no zeros.

Proof. If there exists a solution of (4.1) such that ρ is complex, we know from Floquet's theorem that all solutions are bounded. In this case, formulas (3.18) to (3.20) show that $\phi \to \infty$ as $x \to \infty$ and therefore (4.3) shows that there is a real solution with infinitely many zeros and no real solution with finitely many. Therefore, we know that ρ in (4.4) must be real and that we may take y in (4.4) to be real also, if there exists a solution with finitely many zeros. But in this case, ρ must also be positive since otherwise y would have infinitely many changes of sign, and y itself cannot vanish for any fixed $x = x_0$ because it would also vanish for $x = x_0 + n\pi, n = 0, \pm 1, \pm 2, \ldots$. This proves Lemma 4.1 completely. Now we shall prove

Lemma 4.2. If (4.1) has a real solution with finitely many zeros, then the inequality (4.2) of Theorem 4.2 holds for all continuously differentiable functions $w(x)$ for which $w(x + \pi) = w(x)$.

Proof. We know from Lemma 4.1 that we have a real solution y satisfying (4.3) with a positive ρ. Putting $\rho = \exp \beta$, where β is real, we can write

$$y(x) = e^{\beta x}p(x) = \exp P(x)$$

where

$$p(x + \pi) = p(x), \quad P(x + \pi) = \beta + P(x)$$

and where $P(x)$ is two-times differentiable since $p(x)$ does not vanish. We find easily that

(4.5) $$P'' + (P')^2 + Q^*(x) = 0$$

where $Q^* = Q + \lambda$, and therefore

$$\int_0^\pi Q^*(x)w^2(x)\,dx$$

$$= -\int_0^\pi (P'w^2)'\,dx - \int_0^\pi (P'w - w')^2\,dx + \int_0^\pi (w')^2$$

Since

$$\int_0^\pi (P'w^2)'\,dx = 0$$

because of the periodicity of P' and w, Lemma 4.2 is true. We can now prove

Lemma 4.3. Let λ_0 be the smallest value of λ for which (4.1) has a periodic, nontrivial solution $y = p_0(x)$ with period π. Then for any $\lambda > \lambda_0$, every solution of (4.1) has infinitely many zeros.

Proof. According to Lemma 4.2, it suffices to show that for $\lambda > \lambda_0$ the inequality (4.2) of Theorem 4.2 can be violated by at least one periodic function $w(x)$. We choose $w = p_0(x)$, and we find

(4.8) $$\int_0^\pi [(\lambda + Q)p_0^2 - (p_0')^2]\,dx$$

$$= (\lambda - \lambda_0)\int_0^\pi p_0^2\,dx + \int_0^\pi [(\lambda_0 + Q)p_0^2 - (p_0')^2]\,dx.$$

The second integral on the right-hand side of (4.6) vanishes because

$$-(p_0')^2 + (Q + \lambda_0)p_0^2 = -(p_0 p_0')',$$

and the integral of $(p_0 p_0')'$ from zero to π vanishes since $p_0 p_0'$ is periodic with period π. This proves Lemma 4.3, since the first integral on the right-hand side of (4.6) is positive.

All that remains to do now is an investigation of the case $\lambda \leq \lambda_0$. We shall reduce this problem to the case $\lambda = \lambda_0$ by proving

Lemma 4.4. If (4.1) has a real, nontrivial solution y, with infinitely many zeros, for $\lambda = \lambda^*$, the same is true for all $\lambda \geq \lambda^*$.

Proof. If (4.1) should have a solution with finitely many zeros, we could (according to Lemma 4.1) assume that it has a real solution $y(x)$ of type (4.4) where ρ is positive and $y(x)$ has no zeros. We shall show that this leads to a contradiction. For this purpose, we construct the solution y^* of

$$(y^*)'' + (\lambda^* + Q)y^* = 0$$

which has the same initial conditions at $x = 0$ as $y(x)$. Clearly, y^* is real and, according to our assumption and the remarks made before Lemma 4.1, it will have infinitely many zeros (although it is not identically zero). But then $y(x)$ will also have a zero. For let x_0^* be the smallest positive zero of y^*. We find easily

$$yy^{*\prime} - y^*y' = (\lambda - \lambda^*) \int_0^x y(t)y^*(t)\,dt. \tag{4.7}$$

We may assume that at $x = 0$, both y and y^* are positive. Then, at the first zero x_0 of y^*, clearly $y^{*\prime}$ must be negative since a solution of a homogeneous, linear, second-order differential equation cannot vanish together with its first derivative. Therefore, for $x = x_0$, we have

$$y(x_0)y^{*\prime}(x_0) = (\lambda - \lambda^*) \int_0^{x_0} y(x)y^*(x)\,dx. \tag{4.8}$$

However, if y does not have any zeros, the left-hand side in (2.8) is negative but the right-hand side is positive. Therefore $y(x_0) < 0$ and $y(x)$ has a zero between 0 and x_0. This proves Lemma 4.4.

Now we shall have completed the proof of Theorem 4.1 once we have proved

Lemma 4.5. Let λ_0 and $p_0(x)$ be defined as in Lemma 4.3. Then $p_0(x) \neq 0$ for all x.

The combination of Lemma 4.3 and Lemma 4.5, together with the remarks made before Lemma 4.1, contain the full statement of Theorem 4.1. We shall now prove Lemma 4.5. First, we observe that $p_0(x)$ must have at least two zeros in the half-open interval $0 \leq x < \pi$ if $p_0(x)$ has any zeros at all. This follows from $p(0) = p(\pi)$ and $p'(0) = p'(\pi)$. If $p(0) \neq 0$, $p(x)$ has two zeros in $0 < x < \pi$ since it must change its sign at least twice. If $p(0) = 0$, we have $p'(0) \neq 0$ and if $p'(0) > 0$, it follows that $p(x)$ must be increasing at $x = \pi$. Therefore, $p(\varepsilon) > 0$ and $p(\pi - \varepsilon) < 0$ for a sufficiently small $\varepsilon > 0$.

We now consider

$$\Delta(\lambda_0) = y_1(\pi, \lambda_0) + y_2'(\pi, \lambda_0),$$

where $y_1(x)$ and $y_2(x)$ are the standard solutions of (4.1) for $\lambda = \lambda_0$ defined by the initial conditions (3.17). We know from the general theory of Hill's equation (Part I, Theorem 2.1, Lemmas 2.1 and 2.2) that

$$\Delta(\lambda_0) = 2, \quad \Delta(\lambda) > 2 \quad \text{for} \quad \lambda < \lambda_0.$$

We also know from the proof of Lemma 2.1 in Part I that

$$y_1(x, \lambda^*) > 0, \quad y_2'(x, \lambda^*) > 0$$

for all $x > 0$ if λ^* is such that

$$\lambda^* + Q(x) < 0$$

for all x. We know that $\lambda^* < \lambda_0$ and that

$$\Delta(\lambda^*) > 2.$$

We shall prove Lemma 4.5 by showing that if $p_0(x)$ has at least two zeros in $0 \leq x < \pi$, then there exists a value λ' of λ such that

$$\Delta(\lambda') < 0, \quad \lambda^* < \lambda' < \lambda_0.$$

But in this case, $\Delta(\lambda) - 2$ has a zero between λ^* and λ' which, therefore, is smaller than λ_0 contrary to the fact that, according to Theorem 2.1, Section 2.1, λ_0 is the smallest zero of $\Delta(\lambda) - 2$.

In order to show that λ' exists, we introduce the quantity

$$\psi(\lambda) = \int_0^\pi [y_1^2(x, \lambda) + y_2^2(x, \lambda)]^{-1}\, dx.$$

Then we have according to (3.18) that

$$y_1(x, \lambda) = \rho(x) \cos \phi, \quad y_2(x, \lambda) = \rho(x) \sin \phi,$$

where $\phi(x, \lambda)$ is an increasing function of x such that

$$\phi(0, \lambda) = 0, \quad \phi(\pi, \lambda) = \psi(\lambda).$$

We can write $p_0(x)$ in the form

$$p_0(x) = A\rho \cos (\phi - \phi_0),$$

where, for $0 \leq x \leq \pi$,

$$0 \leq \phi \leq \psi(\lambda_0).$$

Since p_0 is supposed to have two zeros in $0 \leq x < \pi$, we have

$$\psi(\lambda_0) > \pi.$$

On the other hand, since for $\lambda = \lambda^*$ neither y_1 nor y_2 have any zeros for $x > 0$, we see that

$$0 < \psi(\lambda^*) < \pi/2.$$

Since it is clear that $\psi(\lambda)$ depends continuously on λ, it follows that there exists a value $\lambda = \lambda'$ such that

$$\psi(\lambda') = \pi, \quad \lambda^* < \lambda' < \lambda_0.$$

Now we have

$$\begin{aligned} \Delta(\lambda') &= y_1(\pi, \lambda') + y_2'(\pi, \lambda') \\ &= \rho(\pi)[1 + \psi'(\lambda')] \cos \psi(\lambda') + \rho'(\pi) \sin \psi(\lambda') \end{aligned}$$

where $\psi'(\lambda) = d\phi/dx$ for $x = \pi$ and $\rho'(x) = d\rho/dx$. Since $d\phi/dx > 0$, it follows from $\psi(\lambda') = \pi$ that

$$\Delta(\lambda') < 0.$$

This proves our contention and Lemma 4.5. Therefore, Theorem 4.1 has been proved completely.

We can use the method of proving Lemma 4.3 to derive the following

Theorem 4.3. *Let λ_0 be the smallest value of λ for which* (4.1) *has a periodic, nontrivial solution of period π. Let $\lambda < \lambda_0$ and let*

$$y = \exp(\beta x)p(x), \quad p(x + \pi) = p(x), \quad p \not\equiv 0$$

be a Floquet-type solution of (4.1). *Then*

$$\beta \geq (\lambda_0 - \lambda)^{1/2}.$$

Proof. Using the fact that $p(x)$ satisfies the differential equation

$$p'' + 2\beta p' + (\lambda + Q + \beta^2)p = 0,$$

we find by a calculation similar to the one used in proving Lemma 4.3 that for any constant μ,

$$\int_0^\pi [(\mu + Q)p^2 - p'^2]\, dx = (\mu - \lambda - \beta^2) \int_0^\pi p^2\, dx. \tag{4.9}$$

Therefore, we see from Lemma 4.2 that the real, nontrivial solutions of

$$y'' + (\mu + Q)y = 0$$

must all have infinitely many zeros if

$$\mu > \lambda + \beta^2$$

since then the right-hand side in (4.9) is positive. But then it follows from Theorem 4.1 that

$$\lambda + \beta^2 \geq \lambda_0.$$

Finally, we mention the following

Corollary 4.2. *If* $\lambda + Q(x) < 0$ *for all* x, *then the nontrivial solutions of* (4.1) *have only finitely many zeros. If*

$$\int_0^\pi [\lambda + Q(x)]\, dx > 0,$$

the real solutions of (4.1) *will always have infinitely many zeros.*

The proof of the first statement of Corollary 4.1 can be given by using the fact that $y_2(x)$ will have only one zero if $\lambda + Q < 0$ for all x. This follows in the same manner in which Lemma 2.1 of Part I has been proved. The second statement of Corollary 4.1 follows if we use Lemma 4.2 with $w(x) = 1$.

In view of the decisive role which λ_0 plays in the theory of oscillatory solutions it may be appropriate to prove here

Theorem 4.4. *If* $Q(x)$ *is real and satisfies the conditions in* (4.1), *the smallest value* λ_0 *of* λ *for which Hill's equation has a solution of period* π *is not positive, and* $\lambda_0 = 0$ *if and only if* $Q(x)$ *vanishes identically.*

This result was first found by Borg (1946). The very simple proof given here is due to Ungar (1961). A generalization involving λ_{2n} was found by Blumenson (1963) (see Section 5.6, Theorem 5.8). We know from Theorem 4.1 that the periodic solution $p_0(x)$ of period π belonging to $\lambda = \lambda_0$ has finitely many and, therefore, no zeros. We may assume $p_0(x) > 0$ for all x and therefore,

$$h(x) = \frac{d}{dx} \log p_0(x)$$

is a differentiable periodic function of period π satisfying the Riccati equation

$$h'(x) + h^2(x) = -\lambda_0 - Q(x).$$

Integrating this relation from 0 to π yields

$$\int_0^\pi h^2(x)\,dx = -\pi\lambda_0,$$

since the integral over $Q(x)$ vanishes because of (4.1) and the integral over $h'(x)$ vanishes because $h(\pi) = h(0)$. Therefore, λ_0 is negative and can be zero only if the integral over h^2 vanishes. For a continuous h, this means $h \equiv 0$, therefore $p_0(x)$ is a constant. But then $\lambda_0 + Q \equiv 0$, which means $Q \equiv 0$.

V

Intervals of Stability and Instability

5.1. *Introduction*

The general theory, as presented in Part I, has been supplemented by numerous investigations of both special cases and classes of real functions $Q(x)$ for which more precise statements can be obtained. We shall now try to indicate what has been achieved. Since in most cases the results either are complicated or based on numerical computations, we shall confine ourselves to giving a general description of the most important methods together with a few references. Only the results about the so-called "regions of absolute stability" which are due to Borg (1944), will be presented in detail. The notations used will be those of Part I. We shall always write Hill's equation in the form of equation (2.1) as

$$(5.1)\quad y'' + [\lambda + Q(x)]y = 0, \quad Q(x + \pi) = Q(x), \quad \int_0^\pi Q\,dx = 0.$$

and we shall use the terminology of Theorem 2.1. For $Q(x)$ we shall assume a convergent Fourier expansion

$$(5.2)\qquad Q(x) = \sum_{r=-\infty}^{+\infty} g_n e^{2inx}, \quad g_0 = 0, \quad g_{-n} = \bar{g}_n,$$

where a bar denotes the complex conjugate quantity. Unless otherwise stated, we shall assume that

$$(5.3)\qquad \sum_{n=1}^{\infty} n^2|g_n|^2 < \infty.$$

The model for the questions to be asked about Hill's equation has been provided by Mathieu's equation which we shall write in the form

$$y'' + [\lambda - 2\theta \cos 2x]y = 0. \tag{5.4}$$

Thorough accounts of the theory of this equation and its application have been given, at different times, by Strutt (1932), McLachlan (1947), and by Meixner and Schaefke (1954). It may be considered as a two-parameter equation, and the points in the λ, θ plane where the solutions are stable form certain stability regions which have been studied thoroughly. Strutt (1944) has considered numerous other two- and three-parameter equations of Hill's type; we shall describe some of his results in Section 5.3. The intervals of instability of Mathieu's equation (for a fixed θ) are known to decrease exponentially with their order. A generalization of this result found by Hochstadt is formulated as Theorem 2.13.

If we consider Hill's equation as an equation with infinitely many parameters g_r, the dependence of the intervals of instability on these parameters can be described approximately by a theorem due to Erdélyi. This will be discussed in Section 5.4. If we write $Q(x) = \beta\phi(x)$ and subject $\phi(x)$ to certain normalizing conditions, we obtain two-parameter problems for which Borg has given regions of "absolute" stability. These regions will be described in Section 5.2.

The theory of Hill's equation (5.1) is not entirely equivalent to the theory of the equation

$$y'' + \lambda\rho(x)y = 0,$$

where $\rho(x)$ is periodic of period π. This is due to the fact that Liouville's transformation (Section 3.2) may break down even if $\rho(x)$ is positive but not twice differentiable. If we do not restrict $\rho(x)$ to nonnegative functions or if we even admit negative values for $\rho(x)$, the Liouville transformation cannot be applied even if $\rho(x)$ is an analytic function of x. At the end of Section 5.2, we shall briefly describe some generalizations of Liapounoff's theorem which hold even for functions $\rho(x)$ which are not everywhere positive. For an account of the full theory see the monograph by Krein (1951) and Atkinson (1957).

In Section 5.5, we shall describe briefly some results found by an application of the general theory of systems of linear differential equations.

In Section 5.6, we shall review some finer results on the location and length of the intervals of instability which can be obtained if additional assumptions are made about the function $Q(x)$ in (5.1). We may summarize these results by saying that they are contributions to the question of in which way Theorem 2.12 (Section 2.4) states not only the truth (it is correct), but the full truth and nothing but the truth. Theorem 2.12 is an asymptotic formula which implies that for "large" λ the characteristic values λ_n and λ_n' are slightly larger than the squares of certain integers. Sometimes, asymptotic formulas are valid beyond the range of values for which they have been proved. In the present case, however, this is impossible, since Theorem 6.4 shows that, "on the average" these characteristic values are equal to the squares of integers. Therefore, the question arises how large a "large" λ has to be in Theorem 2.12. For the case where $|Q'''(x)|$ is bounded, Hochstadt (1964) has derived a sufficient lower bound for λ. Furthermore, Theorem 2.12 merely states that the nth interval of instability becomes very small as $n \to \infty$. The first results which give [under certain conditions on $Q(x)$] lower bounds for the lengths of the intervals of instability were found by Levy and Keller (1963) and improved by Hochstadt (1964). Apart from these results, we shall also describe another modification of Theorem 2.12 and a generalization of Theorem 4.4.

The question when some intervals of instability will actually disappear can also be answered in certain cases. For this problem, see Chapter VII.

5.2. *Regions of absolute stability* (*Liapounoff-type results*)

We shall now write

$$Q(x) = \beta\psi(x), \tag{5.7}$$

where, according to (5.1)

$$\int_0^\pi \psi(x)\,dx = 0. \tag{5.8}$$

We shall consider all "functions ψ of class p" which are defined by

$$\left\{\pi\int_0^\pi |\psi(x)|^p\,dx\right\}^{1/p} = 1 \tag{5.9}$$

where $p = 1, 2, 3, \ldots,$ or $p = \infty$. If $p = \infty$, (5.9) means, of course, that

$$\max|\psi(x)| = 1.$$

We shall not postulate (5.3) in this section. It suffices to assume that

$\psi(x)$ is continuous except for a finite number of points where $\psi(x)$ may have a jump.

A region in the real λ, β plane will be called a *region of absolute stability for functions of class p*, if, for any point in this region, (5.1) will have stable solutions for all functions $Q = \beta\psi$ where ψ belongs to the class p.

Borg (1944) proved the following

Theorem 5.2. *Let $n = 0, 1, 2, \ldots$. The region of absolute stability for functions $\psi(x)$ of class one is bounded by the curves*

$$\beta_{n+1} = \pm\frac{4(n+1)\sqrt{\lambda}}{\pi}\operatorname{ctg}\left[\frac{\pi\sqrt{\lambda}}{2(n+1)}\right], \quad n^2 < \lambda < (n+1)^2,$$

$$\beta_n = \pm 2\lambda(1 - n/\sqrt{\lambda}), \quad \lambda > 1, \quad n \geq 1,$$

$$\lambda = 0 \quad \text{for} \quad n = 0,$$

and is such that none of these curves is contained in its interior. The open region bounded by these curves is maximal; for any point outside or on the boundary of this region, there exists a function ψ of class one such that not all solutions of (5.1) *are bounded.*

Theorem 5.3. *Let κ be a real variable, $0 \leq \kappa^2 < 1$, and let*

$$K = \int_0^{\pi/2} \frac{ds}{\sqrt{1 - \kappa^2 \sin^2 s}}, \quad E = \int_0^{\pi/2} \sqrt{1 - \kappa^2 \sin^2 s}\, ds.$$

Then the curves defined for $n = 0, 1, 2, \ldots$ by

$$\beta_{n+1} = \pm 8\cdot 3^{-1/2}\pi^{-2}(n+1)^2 K[K^2(\kappa^2 - 1) + 2KE(2 - \kappa^2) - 3E^2]^{1/2},$$

$$\lambda_{n+1} = 4\pi^{-2}(n+1)^2[K^2(\kappa^2 - 1) + 2KE], \quad \lambda > 0,$$

bound the region of absolute stability of the functions of class two. The boundary points do not belong to the region, since for

$$\lambda + Q(x) = 4\pi^{-2}(n+1)^2K^2(1 + \kappa^2) - 8\pi^{-2}(n+1)^2\kappa^2K^2 \operatorname{sn}^2(2(n+1)Kx/\pi)$$

the differential equation (5.1) *has only one periodic solution* (*and, therefore, at least one unbounded solution*).

The periodic solution (*with period π or 2π*) *is*

$$y = \operatorname{sn} t, \quad t = 2(n+1)Kx/\pi,$$

where sn *t is the Jacobian elliptic function with module κ and period $4K$.*

Theorem 5.4. *For the functions of class* ∞, *the region of absolute stability is bounded by the curves*

$$(\lambda_{n+1} + \beta_{n+1})^{1/2} \operatorname{tg} [\pi\sqrt{\lambda_{n+1} + \beta_{n+1}}/4(n + 1)]$$
$$= (\lambda_{n+1} - \beta_{n+1})^{1/2} \operatorname{ctg} [\pi\sqrt{\lambda_{n+1} - \beta_{n+1}}/4(n + 1)],$$

where $n = 0, 1, 2, \ldots$, *and where the region does not contain any one of these curves in its interior. If one of the square roots should be imaginary, the functions* tg *and* ctg *have to be replaced by the corresponding hyperbolic functions.*

Borg also observed that Liapounoff's theorem (Section 2.6) can be derived from Theorem 5.2, and that Theorems 5.3 and 5.4 contain, respectively, the following results as special cases:

If $Q(x) + \lambda > 0$ and

$$\int_0^\pi [\lambda + Q(x)]^2 \, dx < (64/(3\pi^2))\left\{\int_0^{\pi/2} \frac{ds}{\sqrt{1 + \sin^2 s}}\right\}^4,$$

then the solutions of (5.4) are stable. The same is true if the second condition is replaced by

$$\max\,[Q(x) + \lambda] < 1.$$

Finally, Borg indicates how the following result due to Beurling (unpublished) can be derived from his arguments.

Theorem 5.5. *If a and b are real numbers and*

$$a^2 \leq \lambda + Q(x) \leq b^2,$$

then the solutions of (5.1) *will be stable for all possible* $Q(x) + \lambda$ *satisfying this condition if and only if the interval* (a^2, b^2) *does not contain the square of an integer.*

If, instead of (5.1), we consider the differential equation of the vibrating string with mass density ρ:

$$y'' + \lambda\rho(x)y = 0, \quad 0 \leq x \leq l,$$

where ρ is a real, nonnegative function of x which is summable in the interval $(0, l)$, the following situation emerges:

The boundary-value problem $y(0) = y(l) = 0$ has a nontrivial solution for a monotonically increasing sequence of values λ_n of λ. If we put

$$M = \int_0^l \rho(x)\, dx$$

where we may assume $M > 0$, then, for $n = 1, 2, 3 \ldots$,

$$\lambda_n \geq 4n^2/(Ml)$$

These inequalities are best possible in the sense that the equality sign may be correct. This will happen if and only if ρ is defined as follows: Divide the interval $(0, l)$ into n equal parts and put the mass M/n into the center of each of these intervals. In other words, the mass distribution is that of n equally heavy and equally spaced beads. [Of course, ρ is determined only up to a nonnegative function the integral of which over the interval $(0, l)$ vanishes.]

If ρ is restricted by the condition

$$0 \leq \rho(x) \leq H,$$

then there exist upper *and* lower bounds for the λ_n. Putting

$$d = M/H,$$

and defining $\chi(t)$ for real values of t implicitly as the smallest positive root of the equation

$$\sqrt{\chi}\, \mathrm{tg}\, \sqrt{\chi} = t/(1 - t),$$

we have

$$\max \lambda_n = \pi^2 n^2/(Md)$$

$$\min \lambda_n = \frac{4n^2}{Md}\chi\left(\frac{d}{l}\right) > \frac{4n^2}{Ml}\left[1 - \left(1 - \frac{4}{\pi^2}\right)\frac{d}{l}\right]^{-1}$$

There exist functions $\rho(x)$ for both the maximum and for the minimum of λ_n for which λ_n assumes these extremal values. These functions can be determined (up to a nonnegative function with vanishing integral); for the minimal values of λ_n, there exists only one such function ρ. The same is true for the maximum of λ_1 (but not for the maximum of λ_n, $n \geq 2$). For details see Krein (1951) where there can also be found a full discussion of the extremal values of the λ_n in the case where $\rho(x)$ is bounded from below as well as from above, i.e., where

$$0 < h \leq \rho(x) \leq H.$$

If $\rho(x)$ can be continued beyond its original interval of definition as an even, periodic (with period l) differentiable function of x, and if ρ is positive, sufficient conditions exist [Hochstadt (1962)] which guarantee that all solutions of

$$y'' + \rho(x)y = 0$$

are bounded. The conditions are: There exists an integer $k \geq 0$ such that

$$k\pi \leq \int_0^l [\rho(x)]^{1/2}\, dx - \frac{1}{4}\int_0^l \left|\frac{\rho'(x)}{\rho(x)}\right| dx$$

and also

$$(k + 1)\pi \geq \int_0^l [\rho(x)]^{1/2}\, dx + \frac{1}{4}\int_0^l \left|\frac{\rho'(x)}{\rho(x)}\right| dx.$$

Finally, Krein (1951) and Atkinson (1957) discuss the differential equation

$$y'' + \lambda p(x)y = 0$$

where $p(x + l) = p(x)$ and p does not vanish almost everywhere, but where $p(x)$ may assume negative values subject only to the condition that

$$\int_0^l p(x)\, dx \geq 0.$$

In this case the following generalization of Liapounoff's theorem (Section 2.6) holds:

All solutions of $y'' + \lambda py = 0$ *will be bounded (for all* x, $-\infty < x < \infty$) *if*

$$0 < \lambda < \frac{4H}{L^2}\chi\left(\frac{L}{Hl}\right),$$

where H *is such that* $p(x) \leq H$, *where* $\chi(t)$ *has been defined above by a transcendental equation and where*

$$L = \int_0^l p_+(x)\, dx; \quad p_+(x) = \max\,[p(x), 0].$$

In the case where $H = 0$, *the inequalities for* λ *take the form*

$$0 < \lambda < \frac{4}{lL}.$$

For further results and for generalizations see Krein (1951) and Atkinson (1957).

5.3. *Equations with two or more parameters*

Strutt (1944) investigated the differential equations

$$y'' + [\lambda + \gamma\Phi(x)]y = 0 \tag{5.10}$$

and

$$y'' + [\lambda + \gamma_1\Phi_1(x) + \gamma_2\Phi_2(x)]y = 0 \tag{5.11}$$

where Φ, Φ_1, and Φ_2 are real periodic functions of x with period ξ and λ, γ, γ_1, and γ_2 are parameters. His approach is the following one: Let σ be a constant such that $|\sigma| = 1$. Then Strutt asks for the values of λ and γ or [in the case of (5.11)] of λ, γ_1, and γ_2 for which (5.10) or (5.11) has a solution $y = w(x)$ such that

$$w(x + \xi) = \sigma w(x), \quad w'(x + \xi) = \sigma w'(x), \tag{5.12}$$

where w does not vanish identically. In earlier papers [Strutt (1943)] it was demonstrated that, for given values of γ_1 and γ_2, the smallest value of λ for which (5.11) has a solution satisfying (5.12) is the minimum assumed by the expression

$$M = -\int_0^\xi \bar{v}v''\,dx - \gamma_1\int_0^\xi \Phi_1(x)\bar{v}v\,dx - \gamma_2\int_0^\xi \Phi_2\bar{v}v\,dx$$

for the set of all twice-differentiable functions v for which

$$v(x + \xi) = v(x)\sigma, \quad v'(x + \xi) = v'(x)\sigma,$$

and

$$\int_0^\xi v\bar{v}\,dx = 1.$$

Here a bar denotes the complex conjugate quantity. Higher "eigenvalues" λ can be found by maximum–minimum conditions of a more complicated nature. Another approach utilized by Strutt (1944) is the method of linear integral equations. The function $w(x)$ will satisfy the equation

$$w(x) = \int_0^\rho [\gamma_1\Phi_1(t) + \gamma_2\Phi_2(t)]G(x, t)w(t)\,dt, \tag{5.13}$$

when $G(x, t)$ is Green's function for the case $\gamma_1 = \gamma_2 = 0$ which, for

$$\rho\sqrt{\lambda} \neq n\pi, \qquad n = 0, \pm 1, \pm 2, \ldots$$

is given by

$$G(x, t) = \frac{-\sigma\omega^{-1} \sin \omega(x - t + \xi) + \omega^{-1} \sin \omega(x - t)}{\sigma^2 - 2\sigma \cos \omega\xi + 1} \quad \text{for} \quad x \leq t,$$

$$G(x, t) = \frac{\sigma\omega^{-1} \sin \omega(x - t - \xi) - \sigma^2\omega^{-1} \sin \omega(x - t)}{\sigma^2 - 2\sigma \cos \omega\xi + 1} \quad \text{for} \quad x \geq t,$$

where $\omega = \sqrt{\lambda}$. Strutt (1944) derives from (5.13) the inequality

$$\{\gamma_1^2 + \gamma_2^2\}^{-1} \leq \int_0^\xi \int_0^\xi (\Phi_1^2(x) + \Phi_2^2(x))|G(x, t)|^2 \, dx \, dt \tag{5.14}$$

Other results obtained by Strutt refer to the shape of the surfaces in the $\lambda, \gamma_1, \gamma_2$ space belonging to a constant σ, including asymptotic relations.

Strutt applies his results to differential equations arising from the separation of variables of Laplace's equation in four variables in a coordinate system that is based on confocal paraboloids in a subspace of three dimensions. One of the resulting ordinary differential equations arising from separation of variables is

$$y'' + \left\{\eta - \tfrac{1}{8}\alpha^2 - (p + 1)\alpha \cos 2x + \frac{\alpha^2}{8} \cos 4x\right\}y = 0,$$

which is a special case of (5.11). Strutt (1944) discusses the surfaces in the spaces of the parameters belonging to $\sigma = 1$.

5.4. *Remarks on a perturbation method*

In a certain sense, most of the methods dealing with Hill's equation may be considered as perturbation methods. The number of papers applying iteration or perturbation methods to technical or physical problems is rather large, and a good survey of the older literature may be found in Erdélyi (1935). Erdélyi (1934, 1935) investigated closed electric circuits with time dependent, periodic capacitance, inductance, and (Ohmic) resistance, the period being the same for all three quantities. This problem leads (after an application of a Liouville transformation) to an equation of Hill's type, and Erdélyi developed a formal, but rather general, perturbation theory for it, with a critical commentary on the range of validity of the formulas. We shall indicate briefly what may be found in Erdélyi's papers which are particularly systematic and thorough.

In order to make Erdélyi's papers more accessible, we shall use the notation that is used there. Consider

$$\frac{d^2z}{dx^2} + [\lambda - kr(x)]z = 0, \quad r(x + 2\pi) = r(x), \tag{5.15}$$

define λ' by

$$\lambda' = (n/2)^2 - \lambda, \quad n = [\sqrt{2\lambda}],$$

where $[u]$ means the largest integer not exceeding u. Then, for large values of λ, we may write

$$\lambda' = \lambda_0 + (2/n)\lambda_1 + (2/n)^2\lambda_2 + \cdots$$

$$z(x) = z_0(x) + (2/n)z_1(x) + (2/n)^2z_2(x) + \cdots,$$

where

$$z_0(x) = \cos(\tfrac{1}{2}nx - \eta), \qquad \eta = \text{constant},$$

and where

$$z(x) = z_0(x) + \frac{2}{n}\int_0^x [\lambda' + kr(\xi)]z(\xi)\sin\frac{n(x - \xi)}{2}\,d\xi$$

and therefore, for $\nu = 1, 2, 3, \ldots$,

$$z_\nu(x) = \int_0^x [\lambda_0 + kr(\xi)]\sin \tfrac{1}{2}n(x - \xi)\, z_{\nu-1}(\xi)\,d\xi$$
$$+ \sum_{\rho=1}^{\nu-1} \lambda_\rho \int_0^x \sin \tfrac{1}{2}n(x - \xi)\, z_{\nu-\rho+1}(\xi)\,d\xi$$

If λ is a characteristic value, $z(x)$ will be of period 2π or 4π, respectively, if n is even or odd, and conditions for determining η, λ_0, λ_1, λ_2, etc., may be obtained by postulating that $z_\nu(x)$ is periodic with the corresponding period 2π or 4π. Assume that

$$kr(x) = \sum_{\nu=0}^{\infty} \gamma_\nu \cos(\nu x - \varepsilon_\nu).$$

Then the posutlate that $z_1(x)$ shall be periodic leads to the condition

$$\tfrac{1}{2}(\lambda_0 + \gamma_0)\sin(\tfrac{1}{2}nx - \eta) + \tfrac{1}{2}\gamma_n \sin(\tfrac{1}{2}nx + \eta - \varepsilon_n) \equiv 0,$$

which gives, as a first approximation

$$\lambda \approx (\tfrac{1}{2}n)^2 + \gamma_0 + \tfrac{1}{2}\gamma_n \tag{5.16}$$

or

$$\lambda \approx (\tfrac{1}{2}n)^2 + \gamma_0 - \tfrac{1}{2}\gamma_n. \tag{5.17}$$

The solutions belonging to the λ values in (5.16) and (5.17) are, respectively,

$$\cos(\tfrac{1}{2}nx - \tfrac{1}{2}\varepsilon_n), \quad \sin(\tfrac{1}{2}nx - \tfrac{1}{2}\varepsilon_n).$$

Erdélyi continues by giving an approximate form for the unbounded solutions in the case where λ is in an interval of instability, using a method introduced by Whittaker (1914). Critical remarks about the method may be found in Erdélyi (1934), p. 617.

In a subsequent paper [Erdélyi (1935)], the equation

$$y'' + [\lambda + \gamma\Phi(x)]y = 0 \tag{5.18}$$

with

$$\Phi(x + 2\pi) = \Phi(x), \quad \int_0^{2x} \Phi(x)\,dx = 0, \quad |\Phi(x)| \leq 1$$

is treated in the two cases where $|\gamma|$ is small compared to 1 and where λ is positive and large compared to 1, $|\gamma|$ being smaller than λ. Conditions for stability are established and approximate formulas for the solutions are given, in the second case (λ large) by using a refined W.K.B. method.

With respect to the approximations given by (5.16) and (5.17), see also Theorem 6.5.

5.5. *Application of the theory of systems of differential equations*

Haacke (1952), Gambill (1954, 1955), and Golomb (1958) applied their results obtained for systems of linear homogeneous differential equations to the theory of Hill's equation which is a special case of a system of two differential equations of the first order. We shall deal here only with the results of Golomb (1958).

According to Floquet's theorem (see Section 1.2), Hill's equation

$$y'' + Q(x)y = 0 \tag{5.19}$$

always has a pair of solutions $f_1(x)$ and $f_2(x)$ given by

$$f_1(x) = e^{i\alpha x}p_1(x), \quad f_2(x) = e^{-i\alpha x}p_2(x),$$

where $p_1(x)$ and $p_2(x)$ are periodic with period π. (If $\alpha \neq 0, \pm 1, \pm 2, \ldots$, these solutions are linearly independent.) We shall call α and $-\alpha$ the characteristic exponents of (5.19). Assume now that $Q(x)$ is even. Then we may write

$$Q(x) = \omega^2 + 2\theta \sum_{n=1}^{\infty} \gamma_n \cos 2nx. \tag{5.20}$$

Equations (5.19) and (5.20) are equivalent to a system of linear differential equations for two functions w_1 and w_2, defined by

$$w_1 = -\omega y + iy', \quad w_2 = -\omega y - iy',$$

which emerges in the form

$$\frac{\omega}{i}\frac{dw_1}{dx} = \omega^2 w_1 + \theta \sum_{n=1}^{\infty} (\gamma_n \cos 2nx)(w_1 + w_2),$$

$$-\frac{\omega}{i}\frac{dw_2}{dx} = \omega^2 w_2 + \theta \sum_{n=1}^{\infty} (\gamma_n \cos 2nx)(w_1 + w_2).$$

Golomb (1958) shows that α can be determined from the equation

$$\begin{aligned} (5.21) \quad \alpha^2 = \omega^2 &+ \theta^2 \sum_{n=-\infty}^{\infty} \frac{\gamma_{|n|}}{(2n+\alpha)^2 - \omega^2} \\ &+ \theta^3 \sum_{\substack{n,m=-\infty \\ n \neq m}}^{\infty} \frac{\gamma_{|n|}\gamma_{|m|}\gamma_{|n-m|}}{[(2n+\alpha)^2 - \omega^2][(2m+\alpha)^2 - \omega^2]} \\ &+ \mathcal{O}(\theta^4), \end{aligned}$$

(where $\gamma_0 = 0$), provided that θ is sufficiently small. The resulting value for α, apart from an error of the order of θ^4, may be written (with the convention $\gamma_0 = 0$) in the form

$$\begin{aligned} (5.22) \quad \alpha = \pm\Bigg[\omega &+ \frac{\theta^2}{4\omega} \sum_{n=1}^{\infty} \frac{\gamma_n{}^2}{n^2 - \omega^2} \\ &+ \frac{\theta^3}{32\omega} \sum_{\substack{n,m=-\infty \\ n \neq m}}^{\infty} \frac{\gamma_{|n|}\gamma_{|m|}\gamma_{|n-m|}}{nm(n+\omega)(m+\omega)} + \mathcal{O}(\theta^4)\Bigg]. \end{aligned}$$

If, in (5.22), α is equated to 0, 1, 2, . . ., one obtains relations between ω and θ which, if satisfied, guarantee that (5.19) and (5.20) has a periodic solution of period π or 2π.

Formulas for the solutions of (5.19) and answers to equations of convergence were also given by Golomb (1958).

5.6. *The instability intervals*

According to Theorem 2.12 (Section 2.4), the nth interval of instability of (5.1) has end points both of which are to the right of n^2 if n is sufficiently large. But Theorem 6.4 shows that this statement cannot be true for *all n*. Hochstadt (1963c) proved

Theorem 5.6. *Let* $Q(x) = Q(-x)$ *and assume that it satisfies the conditions in* (5.1). *Then the characteristic values* λ'_{2n-1}, λ'_{2n} *and* λ_{2n-1}, λ_{2n} *belonging to Hill's equations* (5.1) *satisfy the inequalities*

$$\lambda'_{2n-1} > (2n-1)^2, \quad \lambda'_{2n} > (2n-1)^2,$$
$$\lambda_{2n-1} > 4n^2, \quad \lambda_{2n} > 4n^2,$$

provided that they are greater than a constant Λ *defined by*

$$\Lambda = \max\left\{M + (1 + \tfrac{1}{4}M)^2, \quad M + \frac{64(a + bM + cM^2)^2}{M^2}\right\},$$

where M *is a uniform bound for* $|Q|, |Q'|, |Q''|, |Q'''|$, *where*

$$a = \frac{1}{16}\left(\frac{27}{23}\right)^2, \quad b = \frac{81}{32\cdot 23}, \quad c = \frac{33}{46}.$$

Of course, Theorem 5.6 is applicable only if Q''' exists and is bounded. The theorem does not claim that the constant Λ is optimal.

Theorem 2.12 also asserts that, if $Q''(x)$ is continuous, then the characteristic values λ'_{2n-1}, λ'_{2n}, λ_{2n-1}, and λ_{2n} are bounded away from the squares n^2 of integers n by a term which does not tend to zero after multiplication by n^2. If $Q(x)$ is more than twice differentiable, this result can be sharpened. We have, according to Jagerman (1962):

Theorem 5.7. *Let* $\lambda \equiv \omega^2$ *and assume that* $Q'''(x)$ *in* (5.1) *is of bounded variation in the interval* $(0, \pi)$. *Define constants* $c_{n,m}$ *and* d *by*

$$c_{n,m} = \int_0^\pi [Q^{(n)}(x)]^m\,dx, \quad d = \int_0^\pi Q(x)[Q'(x)]^2\,dx,$$
$$[n, m = 0, 1, 2, 3, 4; \quad Q^{(0)}(x) = Q(x)]$$

and let

$$A = 2 - 2^{-6}c^2_{0,2}\omega^{-6} + (2^{-6}c_{0,2}c_{0,3} - 2^{-7}c_{0,2}c_{1,2})\omega^{-8}$$

$$B = 2^{-2}c_{0,2}\omega^{-3} + (2^{-4}c_{1,2} - 2^{-3}c_{0,3})\omega^{-5} + (5\cdot 2^{-6}c_{0,4} - 5\cdot 2^{-5}d + 2^{-6}c_{2,2})\omega^{-7}.$$

Then, as $\omega \to \infty$,

$$\Delta(\omega^2) = A \cos \pi\omega + B \sin \pi\omega + \mathcal{O}(|\omega|^{-9}).$$

Since

$$(A^2 + B^2)^{1/2} = 2 + \mathcal{O}(|\omega|^{-9}),$$

we also have

$$\Delta(\omega^2) = 2 \cos \pi\varphi + \mathcal{O}(|\omega|^{-9}),$$

where

$$\pi\varphi = \omega - 2^{-3}c_{0,2}\omega^{-3} + (2^{-4}c_{0,3} - 2^{-5}c_{1,2})\omega^{-5} - (5\cdot 2^{-7}c_{0,4} - 5\cdot 2^{-6}d + 2^{-7}c_{2,2})\omega^{-7}.$$

The result which we formulate now gives an instance in which λ is certainly not "large" in the sense of Theorem 2.12. It also gives a generalization of Theorem 4.4. We have [Blumenson (1963)]

Theorem 5.8. *If $Q'(x)$ exists and is continuous, and if*

$$\int_0^{\pi} Q(x)e^{2ikx}\,dx = 0, \qquad k = 0, 1, 2, \ldots, 2n,$$

then the characteristic value λ_{2n} of (5.1) *satisfies the inequality*

$$\lambda_{2n} \leq 4n^2,$$

and the equality sign holds if and only if $Q(x) \equiv 0$.

The first result of a general nature which provides lower bounds for the lengths of the intervals of instability was found by Levy and Keller (1963), and extended by Hochstadt (1964). We have

Theorem 5.9. *Let $Q(x)$ be a real function of X which has a finite Fourier expansion*

$$Q(x) = \frac{1}{2}\sum_{\nu=-s}^{s} g_\nu e^{2i\nu x}, \quad g_{-\nu} = \bar{g}_\nu$$

where $g_s \neq 0$ and $s > 0$. *Consider Hill's equation*

$$y'' + [\lambda + \beta Q(x)]y = 0,$$

where β is a real parameter. Then, for $\beta \to 0$, the length L_m of the mth interval of instability is given by

$$L_m = \frac{8s^2}{[(p-1)!]^2}\left(\frac{|g_s\beta|}{8s^2}\right)^p + \mathcal{O}(\beta^{p+1}), \quad m = sp,$$

$$L_m = a_m|\beta|^p + \mathcal{O}(\beta^{p+1}), \quad s(p-1) < m \leq sp,$$

where p is an integer (defined by the relations listed which connect s, m, and p) and where a_m is a constant independent of β.

Theorem 5.9 was proved by Levy and Keller for the case where $Q(x) = Q(-x)$. The generalization by Hochstadt to the case where $Q(x)$ need be neither even nor odd is based on a result which states:

Let $\Delta(\lambda, \beta)$ be the discriminant of Hill's equation in Theorem 5.9 (where now the dependency of Δ on β as well as on λ has been made explicit). Then the roots λ_n and λ_n' of the equations

$$\Delta(\lambda, \beta) - 2 = 0, \quad \Delta(\lambda, \beta) + 2 = 0$$

are analytic functions of β for sufficiently small values of $|\beta|$. This is true even if $Q(x)$ is merely real, periodic, and integrable.

VI

Discriminant

The discriminant of Hill's equation,

(6.1) $$y'' + [\lambda + Q(x)]y = 0,$$

where

(6.2) $$Q(x) = \sum_{n=-\infty}^{\infty} g_n e^{2inx}, \qquad g_{-n} = \bar{g}_n$$

has been introduced in Part I, Section 2.2. We shall assume that

(6.3) $$\sum_{n=-\infty}^{\infty} n^2 |g_n|^2 < \infty$$

and that

(6.4) $$g_0 = 0.$$

We shall use the notations introduced in Section 2.2; in particular, we shall define the quantities Δ_n as in (2.44) and (2.45), and we shall use the notation $\sqrt{\lambda} = \omega$ and the definition of the discriminant Δ:

$$\Delta = \sum_{n=0}^{\infty} \Delta_n(\lambda).$$

We recall that the boundary points ($\neq -\infty$)

$$\lambda_0, \lambda_1, \lambda_2, \ldots, \lambda_{2n-1}, \lambda_{2n}, \ldots$$

of the even-numbered intervals of instability are the roots of $\Delta - 2 = 0$ and that the boundary points

$$\lambda_1', \lambda_2', \ldots, \lambda_{2n-1}', \lambda_{2n}', \ldots$$

of the odd-numbered intervals of instability of (6.1) are the roots of

$\Delta + 2 = 0$. For this reason, a more precise and a more explicit knowledge of the Δ_n seems desirable. The first three theorems of this section will provide information of this type. The last two theorems will be of a more theoretical interest in connection with Borg's Theorem 2.12 and with Erdélyi's approximation formulas [see (5.16) and (5.17)]. The method used in proving Theorem 2.4 (of Part I) can be applied to prove the more general

Theorem 6.1. *For real positive* $\lambda \to +\infty$,

$$|\Delta(\lambda) - \sum_{n=0}^{N} \Delta_n(\lambda)| = \mathcal{O}(\lambda^{-(N+1)/2}). \tag{6.5}$$

Theorem 6.1 could also be proved by using the following result which sharpens Theorem 2.5

Theorem 6.2. *Let*

$$c(l_1, l_2, \ldots, l_n)$$

be the coefficient of

$$g_{l_1} g_{l_2} \cdots g_{l_n}$$

in Δ_n. *Then*

$$c(l_1, \ldots, l_n) = A(\lambda) \cos \pi\sqrt{\lambda} + B(\lambda) \sin \pi\sqrt{\lambda}/\sqrt{\lambda}$$

where A and B are rational functions of λ *such that the degree of the denominator exceeds that of the numerator*

$$\text{in } A(\lambda) \text{ by at least } n - \tfrac{1}{2}[n/2]$$
$$\text{in } B(\lambda) \text{ by at least } n - \tfrac{1}{2}[n/2] - \tfrac{1}{2},$$

where $[n/2]$ *is the largest integer not exceeding* $n/2$.

It seems that the only way to prove Theorem 6.2 is the one that uses the definition of Δ as an infinite determinant. Using Theorem 2.9 of Section 2.3, and inspecting the determinant $D_0(\lambda)$, we see that Δ_n can be written as an infinite sum of determinants with n rows and columns, each of which is multiplied by the reciprocal of a polynomial in λ. Instead of writing down the general formula, we shall illustrate the situation by writing down the expressions for Δ_3 and Δ_4 which will make apparent the general law for the formation of Δ_n. We have

(6.6) $-\Delta_3[4\sin^2(\pi\sqrt{\lambda}/2)]^{-1}$

$$= \sum_{l,m=1}^{\infty}\sum_{t=-\infty}^{\infty} \frac{\begin{vmatrix} 0 & g_l & g_{l+m} \\ g_{-l} & 0 & g_m \\ g_{-l-m} & g_{-m} & 0 \end{vmatrix}}{(\lambda-4t^2)[\lambda-4(t+m)^2][\lambda-4(t+l+m)^2]}$$

and

(6.7) $-\Delta_4[4\sin^2(\pi\sqrt{\lambda}/2)]^{-1}$

$$= \sum_{l,m,k=1}^{\infty}\sum_{t=-\infty}^{\infty} \frac{1}{P(\lambda;t,k,l,m)} \begin{vmatrix} 0 & g_l & g_{l+m} & g_{l+m+k} \\ g_{-l} & 0 & g_m & g_{m+k} \\ g_{-l-m} & g_{-m} & 0 & g_k \\ g_{-l-m-k} & g_{-m-k} & g_{-k} & 0 \end{vmatrix}$$

where

(6.8) $$P(\lambda;t,k,l,m) = (\lambda-4t^2)[\lambda-4(t+k)^2] [\lambda-4(t+k+m)^2][\lambda-4(t+k+m+l)^2].$$

We know from Theorem 2.5 the general behavior of the coefficient c of a particular product of degree n in the g_l. If we make ω purely imaginary and $\lambda = \omega^2$ negative, i.e., if we put

$$\omega = i\theta, \quad \lambda = -\theta^2$$

then, for $\theta \to \infty$,

(6.9) $$c(l_1, \ldots, l_n)\left[\sin^2\frac{\pi}{2}\omega\right]^{-1} = \mathcal{O}(\theta^{-d})$$

where d is the smaller of the differences between the degrees of the denominators and the numerators in $A(\omega)$ and $B(\omega)$. (Special attention must be given to the case where these two differences are equal and where d may be greater than either of these two differences because in $A\cos\pi\omega + B\sin\pi\omega$, the asymptotic behavior for $\omega = i\theta$, $\theta \to +\infty$ may be different from that of $A\cos\pi\omega$ and of $B\sin\pi\omega$. This case has to be settled by putting $\omega = (1+i)\theta$, and letting $\theta \to +\infty$.) If we wish to compute d by using (6.9), we may apply the following.

Lemma 6.1. For $\theta \to +\infty$,

(6.10) $$\sum_{m_1,\ldots,m_r=1}^{\infty}\sum_{t=-\infty}^{+\infty}\prod_{\nu=1}^{n}[\theta^2+(t+m_1+\cdots+m_\nu)^2]^{-1}\cdot(\theta^2+t^2)^{-1} = \mathcal{O}(\theta^{r-2n-1}).$$

The proof of (6.10) is based on a comparison of the multiple sum in (6.10) with the multiple integral

$$(6.11)\quad \int_0^\infty \cdots \int_0^\infty d\mu_1 \cdots d\mu_r \int_{-\infty}^\infty d\tau(\theta^2 + \tau^2)^{-1} \prod_{\nu=1}^{n} [\theta^2 + (\tau + \mu_1 + \cdots + \mu_\nu)^2]^{-1}.$$

By substituting $\theta\tau, \theta\mu_1, \cdots, \theta\mu_n$ for $\tau, \mu_1, \cdots, \mu_n$, this integral becomes

$$\theta^{r-2n-1} \int_0^\infty \cdots \int_0^\infty d\mu_1 \cdots d\mu_r \int_{-\infty}^\infty d\tau(1 + \tau^2)^{-1} \prod_{\nu=1}^{n} [1 + (\tau + \mu_1 + \cdots + \mu_\nu)^2]^{-1}$$

and here the integral is a constant independent of θ. On the other hand, it is easily seen that (6.11) is a majorant of the sum in (6.10).

Using Lemma 6.1 we can prove Theorem 6.2 as follows: Looking at the representation of Δ_n as a sum of determinants which are divided by certain products of the type appearing in (6.10) and picking a particular product

$$P_n = g_{l_1} \cdots g_{l_n}$$

appearing in these determinants, we see that P_n can appear in infinitely many of these determinants only if several of the subscripts $l_1, \ldots, l_n$ are coupled in pairs, say l_1, l_2 and l_3, l_4 etc., such that $l_1 = -l_2$, $l_3 = -l_4$, and so on. If we have r such pairs, the coefficient of P_n in Δ_n is essentially an r-fold infinite sum of type (6.10). But in a determinant of the type appearing in (6.7) (which illustrates the case $n = 4$) at most $[n/2]$ such pairs can appear, and this proves Theorem 6.2.

Since a skew-symmetric determinant of odd order always vanishes, we have as a by-product of our determinantal expression for Δ_n

Corollary 6.2. *If $Q(x)$ in (6.1) is an odd function (that is, if $g_{-n} = -g_n$), then $\Delta_n \equiv 0$ for $n = 1, 3, 5, \ldots$.*

As an aid for computational purposes we state now

Theorem 6.3. *For $n = 0, 1, 2, 3, 4$, the values of Δ_n are given by the following formulas:*

$$\Delta_0(\lambda) = 2\cos\pi\sqrt{\lambda}, \quad \Delta_1(\lambda) = 0,$$

$$\Delta_2(\lambda) = \frac{\pi\sin\pi\sqrt{\lambda}}{2\sqrt{\lambda}} \sum_{r=1}^{\infty} \frac{g_r g_{-r}}{\lambda - r^2},$$

$$\Delta_3(\lambda) = \frac{\pi \sin \pi\sqrt{\lambda}}{8\sqrt{\lambda}} \sum_{r,s=1}^{\infty} \frac{(g_r g_s g_{-r-s} + g_{-r} g_{-s} g_{r+s})(r^2 + s^2 + rs - 3\lambda)}{(\lambda - r^2)(\lambda - s^2)[\lambda - (r+s)^2]},$$

$$\begin{aligned}\Delta_4(\lambda) = &-\pi^2 \frac{\cos \pi\sqrt{\lambda}}{16\lambda} \left\{ \sum_{r=1}^{\infty} \frac{g_r g_{-r}}{\lambda - r^2} \right\}^2 \\ &- \frac{\pi \sin \pi\sqrt{\lambda}}{64\sqrt{\lambda}} \sum_{r=1}^{\infty} (g_r g_{-r})^2 \frac{30\lambda^2 - 70\lambda r^2 - 16r^4}{(\lambda - r^2)^3(\lambda - 4r^2)} \\ &- \frac{\pi \sin \pi\sqrt{\lambda}}{\sqrt{\lambda}} \sum_{\substack{r,s=1 \\ r>s}}^{\infty} g_r g_{-r} g_s g_{-s} R_{r,s}(\lambda) \\ &+ \sum_{\substack{k \neq l \\ k,l=1}}^{\infty} \sum_{m=1}^{\infty} (g_{l+m} g_k g_{-l} g_{-k-m} + g_{-l-m} g_{-k} g_l g_{k+m}) S_{k,m,l,}(\lambda) \\ &+ \sum_{k,l,m=1}^{\infty} (g_k g_l g_m g_{-k-l-m} + g_{-k} g_{-l} g_{-m} g_{k+l+m}) S_{k,l,m}(\lambda) \\ &+ \sum_{k,l,m=1}^{\infty} (g_{k+m} g_{l+m} g_{-m} g_{-k-l-m} + g_{-k-m} g_{-l-m} g_m g_{k+l+m}) \\ &\qquad S_{k,l,m,}(\lambda),\end{aligned}$$

where the functions $R_{r,s}(\lambda)$ and $S_{k,l,m}(\lambda)$ are defined by

$$\begin{aligned}R_{r,s}(\lambda) = &\frac{5\lambda^2 - 3\lambda(r^2 + s^2) + r^2 s^2}{\lambda(\lambda - r^2)^2(\lambda - s^2)^2} \\ &+ \frac{10\lambda - 2r^2 - 2s^2}{(\lambda - r^2)(\lambda - s^2)[\lambda - (r-s)^2][\lambda - (r+s)^2]}\end{aligned}$$

and by

$$4S_{k,l,m}(\lambda) = \frac{1}{km}\sigma(k+m, l) - \frac{1}{k(k+m)}\sigma(m, l) - \frac{1}{m(k+m)}\sigma(k, l+m)$$

with

$$\sigma(l, m) = \frac{\pi \sin \pi\sqrt{\lambda}}{8\sqrt{\lambda}} \frac{3\lambda - l^2 - m^2 - lm}{(\lambda - l^2)(\lambda - m^2)[\lambda - (l+m)^2]}.$$

The proof of Theorem 6.3 is based on some juggling with infinite sums. The basic information required is the following: By expanding $\cos (x\omega)/2$ in a Fourier series in the interval $-\pi < x < \pi$ we find

$$\sum_{t=-\infty}^{\infty} \frac{(-1)^t e^{itx}}{\omega^2 - 4t^2} = \frac{\pi \cos (x\omega/2)}{2\omega \sin (\pi\omega/2)}. \tag{6.12}$$

By substituting $-(t + l)$ for t in the left-hand side, the right-hand side of (6.12) does not change, and we find, with $\lambda = \omega^2$,

$$S(l) = \sum_{t=-\infty}^{\infty} \{[\lambda - 4t^2][\lambda - 4(t + l)^2]\}^{-1} \tag{6.13}$$

$$= \frac{1}{2\pi} \int_{-\pi}^{\pi} \sum_{t=-\infty}^{\infty} \frac{(-1)^t e^{itx}}{\omega^2 - 4t^2} \sum_{t=-\infty}^{\infty} \frac{(-1)^t e^{-itx}}{\omega^2 - 4(t + l)^2} dx$$

$$= \frac{\pi^2(-1)^l}{8\pi\omega^2 \sin^2(\pi\omega/2)} \int_{-\pi}^{\pi} \cos^2(x\omega/2) e^{-ilx}\, dx$$

$$= \frac{\pi \cos(\omega\pi/2)}{4\omega \sin(\omega\pi/2)[\omega^2 - l^2]}.$$

Formula (6.13) allows us to compute the value of $\Delta_2(\lambda)$. For the computation of $\Delta_3(\lambda)$ we need

$$S(m, l) = \sum_{t=-\infty}^{\infty} \{[\lambda - 4t^2][\lambda - 4(t + m)^2][\lambda - 4(t + m + l)^2]\}^{-1},$$

and we find after an elementary calculation that

$$S(m, l) = -\frac{S(m)}{4l(m + l)} + \frac{S(l + m)}{4lm} - \frac{S(l)}{4m(m + l)}$$

where $S(l)$ is defined by (6.13).

Finally, we need several sums for the computation of Δ_4, all of which can be obtained through linear combinations of the formulas already used and from their derivatives with respect to λ. For details see Magnus (1959).

The following theorems are of some interest because of their relation to Borg's Theorem 2.12 and to Erdélyi's approximation formulas.

The method used in the proofs was established by Schaefke (1954). We have

Theorem 6.4. *Let the roots of $\Delta(\lambda) + 2$ and of $\Delta(\lambda) - 2$ be denoted as at the beginning of Chapter VI. Then*

$$\sum_{n=1}^{\infty} [\lambda'_{2n-1} + \lambda'_{2n} - 2(2n - 1)^2] = 0 \tag{6.14}$$

$$\lambda_0 + \sum_{n=1}^{\infty} [\lambda_{2n-1} + \lambda_{2n} - 2(2n)^2] = 0. \tag{6.15}$$

Whereas Theorem 2.12 shows that for large n, λ'_{2n-1} and λ'_{2n} exceed

$(2n-1)^2$ and λ_{2n-1} and λ_{2n} exceed $(2n)^2$, it follows from Theorem 6.4 that the same statement cannot be true for all n. Limitations for Theorem 2.12 have been stated in Section 5.6, Theorem 5.6.

The next result refers to the case where $Q(x)$ is an even function of x, i.e., where

$$Q(x) = Q(-x); \quad g_{-n} = g_n \qquad (n = 1, 2, 3, \ldots).$$

In this case, the periodic solutions of (6.1) belonging to the characteristic values λ'_{2n-1} and λ'_{2n} are either even or odd periodic functions of x with period 2π. Let

$$\gamma_1' < \gamma_2' < \cdots < \gamma_n' < \cdots$$

be the ordered sequence of those numbers of the set λ'_{2n-1}, λ'_{2n} to which there belongs an even periodic function of period 2π, and let

$$\sigma_1' < \sigma_2' < \cdots < \sigma_n' < \cdots$$

be the ordered sequence of the remaining numbers of the λ'_{2n-1}, λ'_{2n} to which there belongs an odd periodic solution of (6.1). Similarly, let

$$\gamma_1 < \gamma_2 < \cdots < \gamma_n < \cdots$$

be the characteristic values belonging to the set of numbers λ_{2n-1} and λ_{2n} to which there belong even solutions of (2.1) which are of period π and let

$$\sigma_1 < \sigma_2 < \cdots < \sigma_n < \cdots$$

be the sequence of characteristic values corresponding to odd solutions of period π. Then we have

Theorem 6.5. *The intervals of instability and the Fourier coefficients of Q are related by the equations*

$$\sum_{n=1}^{\infty} (\gamma_n' - \sigma_n') = -2 \sum_{n=1}^{\infty} g_{2n-1} \tag{6.16}$$

$$\lambda_0 + \sum_{n=1}^{\infty} (\gamma_n - \sigma_n) = -2 \sum_{n=1}^{\infty} g_{2n}. \tag{6.17}$$

The following comment should be made concerning Theorem 6.5. For large n,

$$(6.18) \qquad \begin{aligned} |\gamma_n' - \sigma_n'| &= \lambda'_{2n} - \lambda'_{2n-1} \\ |\gamma_n - \sigma_n| &= \lambda_{2n} - \lambda_{2n-1}. \end{aligned}$$

Therefore, the absolute values of the terms in the sums on the left-hand sides of the equations in Theorem 6.5 represent the lengths of the intervals of instability, at least for large values of n. Now Erdélyi (1934) has shown that for sufficiently small values of

$$\sum_{n=1}^{\infty} |g_n|^2,$$

the nth interval of instability is approximately of length $2|g_n|$. Theorem 6.5 shows that, although Erdélyi's result may not be exact for the individual intervals of instability, there exists a weaker but exact substitute for it in the form of a relation between sums.

We shall now prove Theorem 6.4, and we shall confine ourselves to a proof of (6.14). We know from Borg's theorem 2.12 that

$$(2n-1)^2[\lambda'_{2n-1} - (2n-1)^2] \tag{6.19}$$

and

$$(2n-1)^2[\lambda'_{2n} - (2n-1)^2] \tag{6.20}$$

are bounded for $n \to \infty$. We also know that $\Delta + 2$ is a function of order of growth ½ and that therefore for a suitable value of the constant c,

$$\Delta + 2 = c \prod_{n=1}^{\infty} \left(1 - \frac{\lambda}{\lambda'_{2n-1}}\right)\left(1 - \frac{\lambda}{\lambda'_{2n}}\right) \tag{6.21}$$

unless one of the λ'_n vanishes, in which case we would have to use a slight modification of this formula which will not affect the proof. We know from Theorem 6.3 that for large positive values of $\mu = -\lambda$

$$\Delta + 2 = 4\cosh^2 \frac{\pi}{2}\sqrt{\mu} \tag{6.22}$$

$$\left[1 - \frac{\pi \sinh(\pi\sqrt{\mu}/2)}{4\sqrt{\mu}\cosh(\pi\sqrt{\mu}/2)} \sum_{n=1}^{\infty} \frac{|g_n|^2}{\mu + n^2} + \mathcal{O}(\mu^{-5/2})\right],$$

Now consider the behavior of

$$L(\mu) = \frac{d}{d\mu} \log \frac{\Delta + 2}{\cosh^2(\pi\sqrt{\mu}/2)} \tag{6.23}$$

for $\mu \to \infty$.

From the product representation of $\Delta + 2$ and of $\cosh(\pi\sqrt{\mu}/2)$ we find

(6.24)

$$L(\mu) = \sum_{n=1}^{\infty}\left\{\frac{1}{\mu + \lambda'_{2n-1}} + \frac{1}{\mu + \lambda'_{2n}} - \frac{2}{\mu + (2n-1)^2}\right\}$$

$$= -\sum_{n=1}^{\infty}\left\{\frac{\lambda'_{2n-1} - (2n-1)^2}{(\mu + \lambda'_{2n-1})[\mu + (2n-1)^2]} + \frac{\lambda_{2n} - (2n-1)^2}{(\mu + \lambda'_{2n})[\mu + (2n-1)^2]}\right\}$$

$$= -\sum_{n=1}^{\infty}[\lambda'_{2n-1} + \lambda'_{2n} - 2(2n-1)^2]\mu^{-2}$$

$$+ \sum_{n=1}^{\infty}[\lambda'_{2n-1} - (2n-1)^2]\frac{\mu[\lambda'_{2n-1} + (2n-1)^2] + \lambda'_{2n-1}(2n-1)^2}{\mu^2[\mu + (2n-1)^2][\mu + \lambda'_{2n-1}]}$$

$$+ \sum_{n=1}^{\infty}[\lambda'_{2n} - (2n-1)^2]\frac{[\lambda'_{2n} + (2n-1)^2]\mu + \lambda'_{2n}(2n-1)^2}{\mu^2[\mu + (2n-1)^2][\mu + \lambda'_{2n}]}.$$

We wish to show that

$$L(\mu) = \sum_{n=1}^{\infty}[\lambda'_{2n-1} + \lambda'_{2n} - 2(2n-1)^2]\mu^{-2} + \mathcal{O}(\mu^{-5/2}) \tag{6.25}$$

for $\mu \to \infty$. To do this, we have to estimate the last two sums in (6.24) and prove that they are of the order of $\mu^{-5/2}$. Using the boundedness of expression (6.19) we find that

$$\sum_{n=1}^{\infty}[\lambda'_{2n-1} - (2n-1)^2]\frac{\mu[\lambda'_{2n-1} + (2n-1)^2] + \lambda'_{2n-1}(2n-1)^2}{[\mu + (2n-1)^2][\mu + \lambda'_{2n-1}]} \tag{6.26}$$

can be majorized by

$$S = M\sum_{n=1}^{\infty}\frac{\mu + (2n-1)^2}{[\mu + (2n-1)^2]^2}$$

where M is a suitable constant. By using a standard procedure we see that S can be written as

$$S = 2M\int_0^{\infty}\frac{\mu + t^2}{(\mu + t^2)^2}\,dt + \mathcal{O}(\mu^{-1}). \tag{6.27}$$

Now the integral in (6.27) equals $\pi/(2\sqrt{\mu})$ and this proves (6.25).

Next we shall derive an asymptotic expansion for $L(\mu)$ from (6.22).

From some calculations and by using an argument similar to the one we employed in estimating (6.26) we find that

$$L(\mu) = \mathcal{O}(\mu^{-5/2}) \tag{6.28}$$

provided that

$$\sum_{n=1}^{\infty} n^2 |g_n|^2 < \infty.$$

A comparison of (6.25) and (6.28) proves Theorem 6.4.

The proof of Theorem 6.5 may be based on the following remarks: First, we can show that for large n the pair of numbers γ_n', σ_n' is identical with the pair λ'_{2n-1}, λ'_{2n}, apart from the ordering of the numbers. To prove this statement, we will have to consider the equation

$$y'' + [\lambda + \varepsilon g(x)]y = 0 \qquad (0 \leq \varepsilon \leq 1)$$

For $\varepsilon = 0$, $\lambda'_{2n-1} = \lambda'_{2n} = (2n - 1)^2$. If ε increases, the difference $(\lambda'_{2n-1} - \lambda'_{2n})$ will, in general, be different from zero, but it will stay small and one of the periodic solutions belonging to these two numbers will be odd, the other, even. Since both λ'_{2n-1} and λ'_{2n} will not leave a certain neighborhood of $(2n - 1)^2$ bounded by, say, $[(2n - 1)^2 - \frac{1}{2}]$ and $[(2n - 1)^2 + \frac{1}{2}]$, these two numbers will always be the characteristic values belonging, respectively, to the nth even and to the nth odd periodic solution of period 2π, although we do not know whether λ_{2n} belongs to an even or to an odd solution.

From this remark and from Borg's theorem 2.12 we see that the series (6.16) and (6.17) converge absolutely and can be majorized by the series $M \sum n^{-2}$ where M is a constant. Furthermore, we know that the numbers

$$\gamma_n', \quad \sigma_n', \quad \gamma_n, \quad \sigma_n \qquad (\gamma_0 = \lambda_0)$$

are, respectively, the zeros of

$$y_1(\pi/2, \lambda), \quad y_2'(\pi/2, \lambda), \quad y_1'(\pi/2, \lambda), \quad y_2(\pi/2, \lambda)$$

where y_1 and y_2 are the normalized solutions of (6.1) described in Chapter I (see Theorem 1.1). From the method of solving (6.1) by iteration, we find that, for $\lambda = -\mu$, μ large and positive,

$$y_1(\pi/2, -\mu) = \cosh(\pi\sqrt{\mu}/2)\left[1 - \sum_{n=1}^{\infty} \frac{g_{2n-1}}{\mu + (2n-1)^2} + \mathcal{O}(\mu^{-3/2})\right]; \tag{6.29}$$

$$(6.30)\quad y_2'(\pi/2, -\mu) = \cosh(\pi\sqrt{\mu}/2)\left[1 + \sum_{n=1}^{\infty} \frac{g_{2n-1}}{\mu + (2n-1)^2} + \mathcal{O}(\mu^{-3/2})\right];$$

$$(6.31)\quad y_1'(\pi/2, -\mu) = \sqrt{\mu}\sinh(\pi\sqrt{\mu}/2)\left[1 - \sum_{n=1}^{\infty} \frac{g_{2n}}{\mu + 4n^2} + \mathcal{O}(\mu^{-3/2})\right];$$

$$(6.32)\quad y_2(\pi/2, -\mu) = \frac{\sinh \pi\sqrt{\mu}/2}{\sqrt{\mu}}\left[1 + \sum_{n=1}^{\infty} \frac{g_{2n}}{\mu + 4n^2} + \mathcal{O}(\mu^{-3/2})\right],$$

where the $\mathcal{O}(\mu^{-3/2})$ terms may be differentiated with respect to μ, having a derivative of the order of $\mu^{-5/2}$.

Again, the left-hand sides in (6.24) to (6.31) are functions of μ of order of growth ½, and will admit product representations of the type (6.21), e.g., (for $\lambda_0 \neq 0, \gamma_n \neq 0, \sigma_n \neq 0$)

$$(6.33)\qquad y_1'(\pi/2, -\mu) = c(1 + \mu/\lambda_0) \prod_{n=1}^{\infty} (1 + \mu/\gamma_n),$$

$$(6.34)\qquad y_2(\pi/2, -\mu) = c^* \prod_{n=1}^{\infty} (1 + \mu/\sigma_n),$$

where c and c^* are constants. Now we can calculate an asymptotic expansion for

$$\frac{d}{d\mu} \log \frac{y_1'(\pi/2, -\mu)}{y_2(\pi/2, -\mu)}$$

in two different ways, using (6.31) and (6.32), or using (6.33) and (6.34). By equating these two expansions, we find

$$\frac{1}{\mu} + 2\sum_{n=1}^{\infty} \frac{g_{2n}}{[\mu + 4n^2]^2} + \mathcal{O}(\mu^{-5/2}) = \frac{1}{\mu + \lambda_0} + \sum_{n=1}^{\infty} \frac{\sigma_n - \gamma_n}{(\mu + \gamma_n)(\mu + \sigma_n)}$$

or

$$2\sum_{n=1}^{\infty} g_{2n}\mu^{-2} + \mathcal{O}(\mu^{-5/2}) = -\left[\lambda_0 + \sum_{n=1}^{\infty} (\gamma_n - \sigma_n)\right]\mu^{-2} + \mathcal{O}(\mu^{-5/2})$$

which proves (6.17). Equation (6.16) can be proved in the same manner.

VII

Coexistence

7.1. *Introduction*

According to Floquet's theorem (Section 1.2), Hill's equation will, in general, have only one periodic solution (and its constant multiples) of period π or 2π. If it should happen that two linearly independent (and therefore all) solutions of Hill's equation are of period π or 2π, we shall say that two such solutions *coexist* or we shall call this an instance of *coexistence*. It should be noted that, according to the Corollary to Floquet's theorem in Section 1.2, the coexistence problem never arises for solutions of a period $n\pi$ where $n = 3, 4, 5, \ldots$. Also, we should recall here that coexistence of periodic solutions of period π or 2π is equivalent, respectively, to the occurrence of a double root of the equation $\Delta(\lambda) - 2 = 0$ or $\Delta(\lambda) + 2 = 0$. If, say, the nth root of $\Delta(\lambda) = 2$ is a double root, i.e., if

$$\lambda_{2n-1} = \lambda_{2n}, \qquad (n = 1, 2, 3, \ldots)$$

we may say that the $2n$th *interval of instability disappears*; similarly, if

$$\lambda'_{2n-1} = \lambda'_{2n}, \qquad (n = 1, 2, 3, \ldots)$$

we may say that the $(2n - 1)$st interval of instability disappears, and these statements are equivalent with the statement that coexistence of periodic solutions of period π or 2π occurs. It may be useful to recall here that λ_0 can never be a double root of $\Delta - 2 = 0$, and that the zeroth interval of instability from $-\infty$ to λ_0 cannot disappear at all.

It had been noted early and proved repeatedly that for Mathieu's equation

$$y'' + (\lambda + \alpha \cos 2x)y = 0, \tag{7.1}$$

no interval of instability can ever disappear unless $\alpha = 0$, in which case only the zeroth interval of instability remains. For a proof see, e.g., MacLachlan (1947). The methods developed in the following sections would also lead to an easy proof of this fact.

Meissner (1918), studied the equation

$$y'' + \lambda g(x)y = 0 \tag{7.2}$$

where $g(x)$ is a piecewise-constant function assuming two different values in the interval $0 \leq x \leq \pi$. For this case, the coexistence problem can be answered completely; a recent discussion of the situation can be found in a paper by Hochstadt (1961) and also in Section 8.2. However, it is impossible to transform (7.2) into the standard form of Hill's equation used throughout the present monograph, since the Liouville transformation as defined by equations (3.6) to (3.9) cannot be applied to nondifferentiable functions.

For various cases of Hill's equation with an analytic coefficient $Q(x)$ the coexistence problem has been investigated. Of these, the case of Lamé's equation is particularly important; it offers the simplest example of an equation of Hill's type for which all but a finite number k of intervals of instability disappear. In view of the generalized Fourier theorem 2.16, these cases provide particularly simple analogs to the ordinary Fourier theorem, since the integration has to be extended over a finite number of intervals only. Lamé's equation will be discussed in Section 7.3.

It has been shown by Winkler (1958) that all known cases of equations of Hill's type with analytic coefficients and with a decidable coexistence problem are special cases of a four-parametric equation that was called *Ince's equation* by Winkler. It is the most general equation to which Ince's method of three-term recurrence relations can be applied. The theory of Ince's equation will be developed in Section 7.2. The equation can be transformed into the standard form (3.2) by using the substitution (3.3) with a properly defined $A(x)$.

It is necessary to solve a transcendental equation if we wish to decide whether an equation of Hill's type has a periodic solution of period π or 2π. However, once it is known that a given equation of Ince's type has a solution of period π or 2π, in general, merely the solution of a problem of linear algebra is required in order to decide whether *all* solutions of Ince's equation are periodic with the same period. In addition, there exists a very simple necessary condition for the parameters of the equation that must be satisfied if coexistence can occur, regardless of the existence or nonexistence of at least one periodic solution with period π or 2π.

In Sections 7.3 to 7.5, we discuss coexistence for a variety of special equations of Hill's type. In Section 7.6, we report on the cases where all but one or all but two of the intervals of instability disappear and we quote a theorem by Hochstadt (1965), concerning the general case of finitely many intervals of instability.

7.2. *Ince's equation*

We shall discuss the coexistence problem for Ince's equation that will be written in the form

$$(7.3) \qquad (1 + a \cos 2x)y'' + b(\sin 2x)y' + (c + d \cos 2x)y = 0,$$

where a, b, c, and d are real parameters and $|a| < 1$. The transformation

$$y = (1 + a \cos 2x)^{b/(4a)} z$$

carries (7.3) into the equation

$$(7.4) \qquad z'' + \frac{\alpha + \beta \cos 2x + \gamma \cos 4x}{(1 + a \cos 2x)^2} z = 0, \qquad (a \neq 0),$$

where

$$\alpha = c - ab - b^2/8 + ad/2$$
$$\beta = d + ac - b$$
$$\gamma = ad/2 + b^2/8,$$

and (7.4) has the form of Hill's equation. The case where $a = 0$ can be dealt with by the substitution

$$y = w \exp [(b \cos 2x)/4]$$

which leads to

$$(7.5) \qquad w'' + [c - b^2/8 + (d - b) \cos 2x + (b^2/8) \cos 4x]w = 0,$$

and this is the most general equation of Hill's type where the coefficient of w is of the form

$$\alpha + \beta \cos 2x + \gamma \cos 4x$$

with $\gamma \geq 0$.

However, we shall base our discussion on equation (7.3). The first result we shall prove is

Theorem 7.1. *If Ince's equation* (7.3) *has two linearly independent solutions of period* π, *then the polynomial*

$$Q(\mu) = 2a\mu^2 - b\mu - d/2$$

has a zero at one of the points

$$\mu = 0, \pm 1, \pm 2, \ldots.$$

If (7.3) *has two linearly independent solutions of period* 2π, *then*

$$Q^*(\mu) = 2Q(\mu - \tfrac{1}{2}) = a(2\mu - 1)^2 - b(2\mu - 1) - d$$

vanishes for one of the values of $\mu = 0, \pm 1, \pm 2, \ldots.$

To prove Theorem 7.1, we need the following

Lemma 7.1. Let $P(\mu)$ be a polynomial of first or second degree in μ and let D_n $(n = 0, 1, 2, \ldots)$ be elements of a sequence which satisfy the recurrence relations

$$P(n)D_n = P(-n-2)D_{n+1}, \qquad n = 0, 1, 2, \ldots \tag{7.6}$$

and the limit relations

$$\lim_{n \to \infty} n^p D_n = 0 \quad \text{for} \quad p = 1, 2, 3, \ldots. \tag{7.7}$$

Then either all D_n vanish or $P(\mu)$ has an integral root.

Proof. If $P(\mu)$ does not have an integral root then either all of the D_n are zero or none of them vanishes [see (7.6)]. Assume that none of the D_n vanishes and that $P(n)$ and $P(-n-1)$ are different from zero for all integers n. Then we can obtain a contradiction to (7.7) and therefore prove our lemma. To do this, we choose a fixed integer $k > 0$ and derive from (7.6) that

$$D_k = \frac{P(-k-2)P(-k-3)\cdots P(-k-r-2)}{P(k)P(k+1)\cdots P(k+r)} D_{k+r+1} \qquad (r = 0, 1, 2, \ldots). \tag{7.8}$$

Assume that $P(\mu)$ is of second degree and that

$$P(\mu) = A(\mu - \lambda_1)(\mu - \lambda_2).$$

Then (7.8) can be written as

$$(7.9)\quad \frac{\Gamma(k+2+\lambda_1)\Gamma(k+2+\lambda_2)}{\Gamma(k-\lambda_1)\Gamma(k-\lambda_2)} D_k = \frac{\Gamma(k+3+\lambda_1+r)\Gamma(k+3+\lambda_2+r)}{\Gamma(k-\lambda_1+1+r)\Gamma(k-\lambda_2+1+r)} D_{k+r+1},$$

and if $P(\mu)$ is of first degree and

$$P(\mu) = A(\mu - \lambda),$$

then

$$(7.10)\quad \frac{\Gamma(k+2+\lambda)}{\Gamma(k-\lambda)} D_k = (-1)^{r+1} \frac{\Gamma(k+3+\lambda+r)}{\Gamma(k-\lambda+1+r)} D_{k+r+1}.$$

Here Γ denotes the Gamma function. As a simple consequence of Stirling's Formula we have the asymptotic relations

$$(7.11)\quad \begin{aligned} \lim_{t\to+\infty} \left| \frac{\Gamma(t+\rho)}{\Gamma(t)} t^{-\rho} \right| &= 1, \\ \lim_{t\to+\infty} \left| \frac{\Gamma(t+\rho)}{\Gamma(t-\rho)} t^{-2\rho} \right| &= 1. \end{aligned}$$

where ρ is any fixed, real or complex number. Putting $t = k + r + 1$, $\rho = 2\lambda + 2$, we find from (7.10) and (7.7) by letting $r \to \infty$ that

$$(7.12)\quad \frac{\Gamma(k+1+\lambda)}{\Gamma(k-\lambda)} D_k = 0.$$

Similarly, we find that the left-hand side in (7.9) must vanish, and this proves Lemma 7.1. Incidentally, we can use the vanishing of the left-hand sides of (7.9) and (7.10) to prove

Lemma 7.2. If the numbers D_n satisfy the recurrence relations (7.6) and the limit relation (7.7), then $D_n = 0$ for $n > k_0$, where k_0 is the largest nonnegative integer such that $P(k_0) = 0$. If no such integer exists, all the D_n vanish.

We shall apply Lemma 7.2 to the proof of a later theorem. To prove Theorem 7.1, we shall derive now

Lemma 7.3. If Ince's equation has two linearly independent solutions of period π or 2π, then two solutions y_1 and y_2 can be found such that either

$$(7.13)\qquad y_1 = \sum_{n=0}^{\infty} A_{2n} \cos 2nx, \quad y_2 = \sum_{n=1}^{\infty} B_{2n} \sin 2nx,$$

or

$$(7.14)\qquad y_1 = \sum_{n=0}^{\infty} A_{2n+1} \cos (2n+1)x, \quad y_2 = \sum_{n=0}^{\infty} B_{2n+1} \sin (2n+1)x,$$

where, for every positive exponent p,

$$(7.15)\qquad \lim_{n\to\infty} n^p A_{2n} = \lim_{n\to\infty} n^p B_{2n} = \lim_{n\to\infty} n^p A_{2n+1} = \lim_{n\to\infty} n^p B_{2n+1} = 0.$$

Proof. We know from Theorem 2.1 that an equation of Hill's type cannot have a solution of period π and a solution of period 2π (which is not of period π). Since Ince's equation can be transformed into Hill's equation by multiplication of y with a function of period π, the same is true of Ince's equation. Also, we can apply to Ince's equation (7.3) the results of Theorem 1.1, since (7.3) can be transformed into a symmetric equation of Hill's type. Therefore, (7.3) has an even and an odd solution, both with the same period, and if the period is 2π, the solutions must change sign if x is increased by π.

These remarks prove (7.13) and (7.14). To prove (7.15), we observe that the solutions of (7.3) must be analytic in a strip of constant width in the complex x plane that contains the real axis. Therefore, the series in (7.13) and (7.14) must converge for $x = x_1 + ix_2$, where x_1 and x_2 are real, x_1 is arbitrary, and x_2 ranges over a sufficiently small interval defined by

$$|1 + a \cos (x_1 + ix_2)| > 0.$$

Therefore, there exists a constant $M > 1$ such that, for example,

$$\lim_{n\to\infty} M^n A_{2n} = 0,$$

and this fact implies immediately the first limit relation in (7.15). This completes the proof of Lemma 7.3. Now we need

Lemma 7.4. If Ince's equation (7.3) has the solutions y_1 and y_2 defined by (7.13), then the A_{2n} and B_{2n} satisfy the recurrence relations

$$(7.16)\qquad -cA_0 + Q(-1)A_2 = 0,$$

$$(7.17)\qquad Q(n-1)A_{2n-2} + (4n^2 - c)A_{2n} + Q(-n-1)A_{2n+2} = 0,$$

$$(n = 1, 2, 3, \ldots),$$

$$(7.18) \qquad (2^2 - c)B_2 + Q(-2)B_4 = 0,$$

$$(7.19) \qquad Q(n-1)B_{2n-2} + (4n^2 - c)B_{2n} + Q(-n-1)B_{2n+2} = 0, \quad (n = 2, 3, 4, \ldots),$$

and if (7.3) has the solutions defined by (7.14), then the recurrence relations

$$(7.20) \qquad [Q^*(0) - 2(c-1)]A_1 + Q^*(-1)A_3 = 0,$$

$$(7.21) \qquad Q^*(n)A_{2n-1} + 2[(2n+1)^2 - c]A_{2n+1} + Q^*(-n-1)A_{2n+3} = 0,$$

$$(7.22) \qquad [-Q^*(0) - 2(c-1)]B_1 + Q^*(-1)B_3 = 0,$$

$$(7.23) \qquad Q^*(n)B_{2n-1} + 2[(2n+1)^2 - c]B_{2n+1} + Q^*(-n-1)B_{2n+3} = 0,$$

hold for $n = 1, 2, 3, \ldots$.

The proof is obvious if we substitute the series in question into the differential equation. Our last step in proving Theorem 7.1 may be formulated as

Lemma 7.5. Let $D_0, D_1, D_2, \ldots$ and $D_0^*, D_1^*, D_2^*, \ldots$ be defined, respectively, by

$$D_0 = A_0B_2, \quad D_n = -A_{2n+2}B_{2n} + B_{2n+2}A_{2n}, \qquad (n = 1, 2, 3, \ldots)$$

and

$$D_0^* = A_1B_1, \quad D_n^* = -A_{2n+1}B_{2n-1} + A_{2n-1}B_{2n+1}, \quad (n = 1, 2, 3, \ldots).$$

Then the following recurrence relations hold:

$$(7.24) \qquad Q(n)D_n = Q(-n-2)D_{n+1}, \qquad (n = 0, 1, 2, \ldots),$$

$$(7.25) \qquad 2Q^*(0)D_0^* = Q^*(-1)D_1^*,$$

$$(7.26) \qquad Q^*(n)D_n^* = Q^*(-n-1)D_{n+1}, \qquad (n = 1, 2, 3, \ldots).$$

The proof is immediate from the recurrence relations of Lemma 7.4.

Now the proof of Theorem 7.1 is easy. Lemma 7.1 and 7.3 show that all D_n of Lemma 7.5 must vanish if $Q(\mu)$ does not have a root which is an integer. But then $D_0 = A_0B_2 = 0$ and either $B_2 = 0$ or $A_0 = 0$. If $B_2 = 0$, the recurrence relations of Lemma 7.4 show that all B_{2n} must vanish [since $Q(n-1) \neq 0$], and therefore the series y_2 in (7.13) is identically zero and y_1 and y_2 are not linearly independent. If $A_0 = 0$, it follows that y_1 vanishes identically.

Since the proof of Lemma 7.1 would also go through if, in (7.6), we replace $P(-n-2)$ by $P(-n-1)$, it follows from a modified Lemma 7.1 and from Lemma 7.3 that all the D_n^* in Lemma 7.5 must vanish when $Q^*(\mu)$ has no integral root. Again $D_0^* = A_1B_1 = 0$ implies that either all A_{2n+1} or all B_{2n+1} vanish which, in turn, would contradict Lemma 7.3. This completes the proof of Theorem 7.1.

We have now a necessary condition for coexistence of two periodic solutions of (7.3) with periods π or 2π, and this condition can be stated even without knowing whether (7.3) has at least *one* periodic solution or not. In order to obtain sufficient conditions for the coexistence of periodic solutions of Ince's equation we have to assume that the values of the parameters are such that at least one solution with period π or 2π exists. We shall settle here first the question of the existence of such values of the parameters by proving

Theorem 7.2. *For any given real values of the parameters a, b, and d (with $|a| < 1$), there exist infinitely many values of the parameter c such that Ince's equation has an even or an odd periodic solution of period π or 2π.*

Before proving Theorem 7.2 we note that the parameter c does not enter into the necessary conditions (as stated in Theorem 7.1) for Ince's equation to have periodic solutions of period π or 2π. To prove Theorem 7.2, we shall write (7.4) in the form

$$(7.27)\qquad z'' + \left[\frac{c}{1 + a\cos 2x} + \frac{\rho + \sigma\cos 2x + \tau\cos 4x}{(1 + a\cos 2x)^2}\right]z = 0,$$

where ρ, σ, and τ do not depend on c. Let us denote the coefficient of z by $H(x)$, and let λ^2 be the minimum and Λ^2 the maximum of H in $0 \le x \le \pi$. Obviously, both λ^2 and Λ^2 increase as c increases, and they tend to infinity as $c \to +\infty$. Let z_1 and z_2 be the standard solutions of (7.27) defined by $z_1(0) = z_2'(0) = 1$ and $z_1'(0) = z_2(0) = 0$. The number of zeros of z_1, z_2, z_1', and z_2' in the interval $0 \le x \le \pi/2$ is then majorized or minorized, respectively, by the number of zeros of the corresponding solutions of $z'' + \Lambda^2 z = 0$ or $z'' + \lambda^2 z = 0$. Since the zeros of the solutions depend continuously on c, it follows that, with increasing c, we must obtain infinitely many values for c such that, for the solutions of (7.27), one of the quantities

$$z_1(\pi/2),\quad z_1'(\pi/2),\quad z_2(\pi/2),\quad z_2'(\pi/2)$$

vanishes. Now an application of Theorem 1.1 will prove Theorem 7.2.

We shall now try to find out under which circumstances the necessary conditions of Theorem 7.1 for coexistence are also sufficient. We shall need the following

Definition. A solution of Ince's equation (7.3) that is given by a series of type (7.13) is called *finite of order k* if A_{2k} or B_{2k} is different from zero, and all A_{2n} or B_{2n} vanish with $n > k$. Similarly, y_1 (or y_2) in (7.14) will be called finite of order k if A_{2k+1} (or B_{2k+1}) is $\neq 0$, but $A_{2n+1} = 0$ (or $B_{2n+1} = 0$) for $n > k$.

Our results will be stated by formulating several theorems.

Theorem 7.3. *If $Q(\mu)$, as defined in Theorem 7.1, has a nonnegative integral root, and if k_0 is the largest such root, then Ince's equation will have two linearly independent solutions of period π provided that one such solution [of type (7.13)] exists that is either infinite or finite of an order $k > k_0$. Similarly, two linearly independent solutions of period 2π will exist if $Q^*(\mu)$ has an integral nonnegative root $k_0{}^*$ (and no larger one), provided that one solution of type (7.14) exists that is infinite or of finite order $k^* > k_0{}^*$.*

Proof. Consider the case where Ince's equation has a solution

$$y_2 = \sum_{n=1}^{\infty} B_{2n} \sin 2nx$$

such that $B_{2k} \neq 0$ for a $k > k_0$. In general, we shall define the solution

$$y_1 = \sum_{n=0}^{\infty} A_{2n} \cos 2nx$$

in the following manner: Let $A_{2n} = B_{2n}$ for $n > k_0$. Our assumptions then guarantee that y_1 is not identically zero and that the series defining it converges everywhere. If y_1 is to be a solution of (7.3), the recurrence relations (7.16) and (7.17) must be satisfied. For $n > k_0$ this will be automatically true since the B_{2n} satisfy the same recurrence relations as the A_{2n}. Note that this is true even if $k_0 = 0$, because then (7.18) has the same shape as (7.17) for $n = 1$. For $n \leq k_0$, we must determine the A_{2n} from the following system of linear equations

(7.28) $$-cA_0 + Q(-1)A_2 = 0,$$

and, if $k_0 > 0$,

(7.29) $$Q(n-1)A_{2n-2} + (4n^2 - c)A_{2n} + Q(-n-1)A_{2n+2} = 0, \quad (n = 1, \ldots, k_0 - 1),$$

and finally

(7.30) $$Q(k_0 - 1)A_{2k_0-2} + (4k_0^2 - c)A_{2k_0} = -Q(-k_0 - 1)B_{2k_0+2}.$$

There are exactly $k_0 + 1$ linear equations for the unknowns $A_0, \ldots, A_{2k_0}$. In general, they will have a unique solution, namely when the determinant of the matrix K of the coefficients of the unknowns is different from zero. In this case we have constructed y_1 in the required manner. However, it may happen that K has determinant equal to zero. Then we shall define y_1 as follows: Determine $A_0, \ldots, A_{2k_0}$ as a nontrivial solution of the homogeneous equations (7.28), (7.29), and (7.30) (replacing B_{2k_0+2} by zero). For $n > k_0$, put $A_{2n} = 0$. Since $Q(k_0) = 0$, this leads to a nontrivial solution of our system of recurrence relations and to a solution y_1 of finite order $\leq k_0$.

A similar argument can be used to prove Theorem 7.3 in the remaining three cases where another one of the four series in (7.13) and (7.14) is assumed to be a solution of (7.3). The case where $Q(0) = 0$ and a series of type y_1 in (7.13) is a given periodic solution requires special attention but is perfectly trivial.

Having disposed of the case where Q or Q^* has a nonnegative integral zero, we can now discuss the situation described in the following

Theorem 7.4. *Assume that $Q(\mu)$ (as defined in Theorem 7.1) has one or two negative integral zeros but none that is ≥ 0. Let $(-k' - 1)$ denote this zero or one of them if there are two, where $k' = 0, 1, 2, \ldots$. Then Ince's equation will never have a solution of period π that is of finite order. If it has one (infinite) solution y of period π, it will have two linearly independent ones if and only if the coefficient of* $\cos 2k'x$ *(for an even y) or of* $\sin 2k'x$ *(for an odd y) in the expansion* (7.13) *vanishes. In this case all Fourier coefficients of the periodic solution with an index less than $2k'$ also vanish.*

Proof. Since $Q(\mu) \neq 0$ for $\mu = 0, 1, 2, \ldots$, it follows from the recurrence relations (7.16) to (7.19) that all the A_{2n} (or all the B_{2n}) vanish for $n \leq l$ if A_{2l+2} and A_{2l+4} (or B_{2l+2} and B_{2l+4}) vanish. Therefore, periodic solutions of finite order cannot exist. Assume now that (7.3) has solutions y_1 and y_2 of type (7.13). According to Lemma 7.2, all the D_n vanish. In particular, $A_0B_2 = 0$. Assume $A_0 = 0$, and let A_{2l} be the first one of the A_{2n} that does not vanish. A glance at (7.17) shows that then $l = k' + 1$, where $Q(-k' - 1) = 0$. Since, according to Lemma 7.2,

$$A_{2l-2}B_{2l} = A_{2l}B_{2l-2},$$

and since $A_{2l-2} = 0$, it follows that $B_{2l-2} = 0$. A similar argument is valid if $B_2 = 0$. Therefore, the conditions for coexistence stated in Theorem 7.4 are necessary ones. But they are also sufficient. For suppose again that

$$A_0 = A_2 = \cdots = A_{2l-2} = 0, \quad B_{2l-2} = 0,$$

and that (7.3) has a solution y_2 of type (7.13). Then we may choose y_1 to be defined by

$$y_1 = \sum_{n=k'+1}^{\infty} B_{2n} \cos 2nx,$$

and the Fourier coefficients of y_1 will satisfy (7.16) and (7.17), since those of y_2 satisfy (by assumption) (7.18) and (7.19). We see easily that $B_{2l-2} = 0$ implies that the previous B_{2n} also vanish.

A result similar to that of Theorem 7.4 holds for solutions of period 2π. We have

Theorem 7.4*. *Assume that $Q^*(\mu)$ (as defined in Theorem 7.1) has one or two negative integral roots but none that is ≥ 0. Let $(-k^* - 1)$ denote one of these roots, where $k^* = 0, 1, 2, \ldots$. Then Ince's equation will never have a solution of finite order that is of period 2π. If it has at least one (infinite) solution y of period 2π, it will have two such solutions that are linearly independent if and only if the coefficient of $\cos(2k^* + 1)x$ (if y is even) or of $\sin(2k^* + 1)x$ (if y is odd) in the Fourier expansion of y vanishes.*

The proof follows exactly the line of the proof of Theorem 7.4 and will be omitted here.

Theorems 7.1 to 7.4* allow us to decide for (7.3) whether two linearly independent solutions of period π or 2π exist provided that one such solution is known, unless the known solution is of finite order and there do not exist two linearly independent solutions of finite order.

Because of the importance of solutions of finite order for the coexistence problem we shall prove

Theorem 7.5. *A necessary condition for Ince's equation to have two linearly independent solutions of finite order is that $Q(\mu)$ (for period π) or $Q^*(\mu)$ (for period 2π) has two integral roots, at least one of which is positive. The order of the finite solutions cannot exceed that largest positive root of $Q(\mu)$ or $Q^*(\mu)$.*

Proof. As in the proof of Theorem 7.4 it follows from the recurrence relations (7.16) to (7.22) that A_{2l} vanishes if A_{2l+2} and A_{2l+4} vanish, unless $Q(2l) = 0$. Similar arguments hold for the three other solutions of type (7.13) and (7.14), and therefore a solution of finite order can exist only if $Q(\mu)$ or $Q^*(\mu)$ has a nonnegative root, and the order cannot exceed the largest root of this type. An inspection of the recurrence relations shows immediately that at most one solution of finite order can exist if $\mu = 0$ is the largest integral root of $Q(\mu)$ or of $Q(\mu^*)$. Therefore $Q(\mu)$ or $Q^*(\mu)$ must have at least one positive integral root. The rest of Theorem 7.5 will be proven if we can show that $b/(2a)$ must be an integer. Because in this case,

$$\mu^2 - \beta\mu - d/(4a) = 0, \qquad (\beta = b/(2a)),$$

must have an integral root. Since β is integral, it follows that $d/(4a)$ is also an integer, and therefore $Q(\mu)$ has two integral roots. A similar argument works for $Q^*(\mu)$. Therefore, all that remains to be shown is

Lemma 7.6. If (7.3) has two linearly independent solutions of finite order, then $b/(2a)$ is a nonnegative integer.

Proof. If y_1 and y_2 are solutions of finite order of (7.3), then

$$w = y_1y_2' - y_2y_1'$$

is also of finite order. If y_1 and y_2 are linearly independent, w does not vanish identically. Also

$$(1 + a\cos 2x)w' + b(\sin 2x)w = 0,$$

and therefore we have, with a constant $w_0 \neq 0$,

$$w = w_0(1 + a\cos 2x)^{b/(2a)}.$$

Obviously, w will have a finite Fourier expansion if and only if $b/(2a) = 0, 1, 2, 3, \ldots$.

In our discussion of Ince's equation, none of the parameters a, b, c, and d plays exactly the role of the eigenvalue parameter λ of the standard Hill's equation used in Chapter I. However, in special cases c may play this role, and we then can talk about intervals of stability and instability. The following result prepares the way for the investigations of this type in later sections. We have

Theorem 7.6. *Let a, b, and d be such that* $Q(\mu)$ *or (for period* 2π*)* $Q^*(\mu)$ *has an integral nonnegative root k or (for period* 2π*)* k^*. *If there are two such solutions, k or* k^* *shall denote the larger one of the two. Let C be the (infinite) set of all values of c for which, with the given values of a, b, and d, (7.3) has at least one periodic solution of period* π *or* 2π. *Then there will be at most* $k + 1$ *or (for period* 2π*)* $k^* + 1$ *values* $c_0, \ldots, c_k$ *or* $c_0^*, \ldots, c_{k^*}^*$ *in C for which (7.3) has a solution of finite order. For all other values of* $c \in C$, *Ince's equation will have two linearly independent solutions of period* π *or* 2π.

Proof. According to Theorem 7.3, Ince's equation will always have two linearly independent solutions of period π or 2π, if the assumptions of Theorem 7.6 are satisfied and if one solution of period π or 2π is given that is of infinite order. According to Theorem 7.3, the order of a finite solution cannot exceed k or k^*. All we have to show now is that there are at most $k + 1$ (or $k^* + 1$) different values of c for which such a finite solution can exist. By inspecting the recurrence relations (7.16) to (7.22) we see that the existence of such a finite solution is equivalent to the existence of a nontrivial solution for a system of at most $k + 1$ (or $k^* + 1$) linear homogeneous equations with an equal number of unknowns.

In turn, the existence of such a nontrivial solution is equivalent to the vanishing of the determinant of the system. Since every row of the determinant in question involves c linearly, the determinant is a polynomial on c of a degree $\leq k + 1$ (or $k^* + 1$) and does not have more zeros than its degree indicates. This proves Theorem 7.6.

We can prove an analog to Theorem 7.6 which shows that, in general, coexistence takes place if $Q(\mu)$ or $Q^*(\mu)$ has a negative integral root. We have

***Theorem* 7.7.** *Let a, b, and d be such that $Q(\mu)$ or (for period 2π) $Q^*(\mu)$ has a negative integral root. Let $(-k_0 - 1)$ or [for roots of $Q^*(\mu)$], $(-k_0^* - 1)$ be the smallest of these roots. Then there exist at most $k_0 + 1$ or (for period 2π) $k_0^* + 1$ values of c such that Ince's equation has one periodic solution of period π or 2π but not two linearly independent ones.*

Proof. We observe that in the proof of Theorem 7.4, $B_2, \ldots, B_{2l-2}$ will satisfy a homogeneous system of linear equations. This can have a nontrivial solution for only a finite number of values of c, as may be seen by the argument used in proving Theorem 7.3.

7.3. Lamé's equation and generalizations

Lamé's differential equation may be written in the form

$$y'' + [\lambda - m(m + 1)k^2 \operatorname{sn}^2 x]y = 0, \tag{7.31}$$

where sn x is Jacobi's elliptic function defined by

$$x = \int_0^{\operatorname{sn} x} \left[(1 - t^2)(1 - k^2t^2)\right]^{-1/2} dt. \tag{7.32}$$

Here k^2 is called the module of sn x. The basic periods of sn x are denoted by $2K$ and $2iK'$; we are only interested in the real period $2K$ which is given by

$$K = \int_0^{\pi/2} \frac{d\phi}{\sqrt{1 - k^2 \sin^2 \phi}}. \tag{7.33}$$

Lamé's equation arises from the theory of the potential of an ellipsoid. It is discussed in detail in Chapter 15 of Erdélyi, et al. (1955). Obviously, Lamé's equation is of the standard type introduced for Hill's equation in Chapter I. We can apply the terminology used there and, in particular, we can talk about intervals of instability on the λ axis. Our main result will be the following

***Theorem* 7.8.** *If and only if m is an integer can periodic solutions of period 2K or 4K of Lamé's equation coexist. If l is defined by $l = m$ if m is a nonnegative integer and by $l = -m - 1$ if m is a negative integer, then Lamé's equation will have at most $l + 1$ intervals of instability (including the zeroth interval which starts at $\lambda = -\infty$).*

Proof. We shall transform (7.31) into an equation of Ince's type by substituting

$$t = \operatorname{am}(x, k), \tag{7.34}$$

where the function am (x, k) is defined by

$$\frac{dt}{dx} = (1 - k^2 \sin^2 t)^{1/2} \qquad (k^2 < 1).$$

Then Lamé's equation takes the form

$$(1 - k^2 \sin^2 t)\frac{d^2y}{dt^2} - \tfrac{1}{2}k^2(\sin 2t)\frac{dy}{dt} + [\lambda - m(m + 1)k^2 \sin^2 t]y = 0, \tag{7.35}$$

which is Ince's equation with

$$a = -b = k^2/(2 - k^2),$$
$$c = [2\lambda - m(m + 1)k^2]/(2 - k^2),$$
$$d = m(m + 1)k^2/(2 - k^2).$$

The period $2K$ of (7.31) corresponds to π for (7.35). The roots of the polynomial $Q(\mu)$ belonging to (7.35) are

$$\mu_1 = m/2, \quad \mu_2 = -(m + 1)/2$$

and the roots of $Q^*(\mu)$ are

$$\mu_1^* = (m + 1)/2, \quad u_2^* = -m/2.$$

None of the numbers $\mu_1, \mu_2, \mu_1^*, \mu_2^*$ can be an integer unless m is an integer. We may assume that $m \geq 0$, since Lamé's equation remains unchanged if m is replaced by $-m - 1$. If m is an even integer, $m = 2l'$, then

$$\mu_1 = l', \quad \mu_2^* = -l'.$$

According to Theorem 7.6, there will not exist more than $l + 1$ values of c (and, therefore, of λ) for which one but not two linearly independent solutions of period π exist. Similarly, it follows from Theorem 7.7 that not more than l' values of c and of λ exist for which (7.35) has one but not two linearly independent solutions of period 2π. Therefore, at most $2l' + 1 = m + 1$ intervals of instability may remain. The case where m is odd can be dealt with in the same manner.

Erdélyi (1941) showed also that not fewer than $m + 1$ intervals of instability remain if the parameter m in Lamé's equation is a non-negative integer.

There are several generalizations of Lamé's equation which can be treated in the same manner. We shall list here the results briefly, following Winkler (1958):

The Hermite elliptic equation can be written in the form

$$(7.36) \quad y'' + \frac{2(r+1)k^2 \operatorname{sn} x \operatorname{cn} x}{\operatorname{dn} x} y' + [\lambda - (m-r)(m+r+1)k^2 \operatorname{sn}^2 x]y = 0,$$

where k^2 and $\operatorname{sn} x$ are defined as above and

$$\operatorname{cn} x = (1 - \operatorname{sn}^2 x)^{1/2}, \quad \operatorname{dn} x = (1 - k^2 \operatorname{sn}^2 x)^{1/2}.$$

The transformation $t = \operatorname{am}(x, k)$ carries (7.36) into

$$(7.37) \quad (1 - k^2 \sin^2 t) \frac{d^2y}{dt^2} + [(2r+1)k^2 \sin t \cos t] \frac{dy}{dt} + [\lambda - (m-r)(m+r+1)k^2 \sin^2 t]y = 0.$$

This is Ince's equation (7.3) with

$$\begin{aligned} a &= k^2/(2 - k^2), \\ b &= (2r+1)k^2/(2 - k^2), \\ c &= [2\lambda - (m-r)(m+r+1)k^2]/(2 - k^2), \\ d &= (m-r)(m+r+1)k^2/(2 - k^2). \end{aligned}$$

The roots of $Q(\mu)$ are given in this case by

$$\mu_1 = (r + m + 1)/2, \quad \mu_2 = (r - m)/2,$$

and the roots of $Q^*(\mu)$ are given by

$$\mu_1{}^* = [(m + r)/2] + 1, \quad \mu_2{}^* = (r - m + 1)/2,$$

A necessary and sufficient condition for one of them to be an integer is that $r + m$ or $r - m$ be an integer.

The *Picard elliptic equation* may be written in the form

$$y'' + \frac{2(r+1)k^2 \operatorname{sn} x \operatorname{cn} x}{\operatorname{dn} x} y' + \lambda y = 0.$$

The transformation $t = \mathrm{am}\,(x, k)$ carries it into Ince's equation with

$$a = k^2/(2 - k^2), \quad b = (2r + 1)k^2/(2 - k^2)$$
$$c = 2\lambda/(2 - k^2), \quad d = 0.$$

In this case, $Q(\mu)$ always has the root zero and therefore all the even intervals of instability except for the zeroth vanish. Odd intervals of instability will vanish if and only if r is an integer.

Equation (7.36) can be transformed into an equation of the form

$$u'' + [H + p\,\mathrm{sn}^2\,x + q(\mathrm{cn}^2\,x/\mathrm{dn}^2\,x)u = 0$$

with constant H, p, and q by putting

$$y = u(\mathrm{dn}\,x)^{-r-1}.$$

This equation for u was called the "Associated Lamé Equation" by Ince.

7.4. *The Whittaker-Hill equation*

The differential equation

$$y'' + [\lambda + 4mq\cos 2x + 2q^2\cos 4x]y = 0 \tag{7.39}$$

appeared first in the work of Liapounoff (1902). It was studied extensively by Whittaker (1915), in a paper on differential equations whose solutions satisfy homogeneous integral equations. Whittaker remarked that (7.39) is related to the Mathieu equation in the same manner in which the equation for the associated Legendre functions is related to Bessel's equation. Ince (1923) pointed out that Whittaker's equation (7.39) has the same relation to the confluent hypergeometric equation as the Mathieu equation has to the Bessel equation. Ince (1926) discussed the real zeros of the solution of Whittaker's equations. Klotter and Kotowski (1943) conducted extensive numerical calculations in connection with (7.39) which resulted in a stability chart and examples for coexistence of periodic solutions. Humbert (1926a) obtained the Whittaker equation from a separation of the Laplace equation (in four dimensions) by introducing hypercylinders formed by three-dimensional confocal paraboloids.

The Whittaker equation (7.39) is not the most general Hill's equation with three real parameters. The assumption that q be real forces the coefficient of $\cos 4x$ to be nonnegative. It has been shown by Winkler

(1958) that coexistence of periodic solutions of (7.39) cannot occur if the coefficient of $\cos 4x$ is replaced by $-2q^2$ (and λ, m, and q are real).

The substitution

$$y = u \exp (q \cos 2x)$$

carries (7.39) into a special case of Ince's equation, namely

$$(7.40) \quad u'' - 4q(\sin 2x)u' + [\lambda + 2q^2 + 4(m - 1)q \operatorname{cox} 2x]u = 0.$$

The polynomials $Q(\mu)$ and $Q^*(\mu)$, as defined in the theory of Ince's equation (Section 7.2), are linear for (7.40); we have

$$Q(\mu) = 4q\mu - 2(m - 1)q;$$
$$Q^*(\mu) = 4q(2\mu - 1) - 4(m - 1)q.$$

If $q \neq 0$ (the case $q = 0$ is trivial), we see that $Q(\mu)$ has the root

$$\mu = (m - 1)/2,$$

and $Q^*(\mu)$ has the root

$$\mu^* = m/2.$$

Now the theory of Ince's equations gives us the following

Theorem 7.9. *Whittaker's equation* (7.39) *can have two linearly independent solutions of period π or 2π if and only if m is an integer. If $m = 2l$ is even, then the odd intervals of instability on the λ axis disappear, with at most $|l| + 1$ exceptions, but no even interval of instability disappears. If $m = 2l + 1$ is odd, then at most $|l| + 1$ even intervals of instability remain but no odd interval of instability disappears.*

7.5. *Finite Hill equations*

The solutions of the Ince equation in the case of coexistence are such that the coefficients of the sine series and the coefficients of the cosine series are equal except for at most the first N coefficients, which sometimes all vanish identically. However, for the Whittaker equation, as a result of the transformation required to put it into the form of an Ince equation, these Ince-type solutions do not occur. Instead, the solutions of the Whittaker equation are Ince type multiplied by an exponential in $\cos 2x$. That Ince-type solutions cannot coexist for a finite Hill equation is the content of the following

Theorem 7.10. *In the differential equation*

$$(7.41) \qquad y'' + (C_0 + C_2 \cos 2x + \cdots + C_{2m} \cos 2mx)y = 0,$$

where $m > 0$ is any integer, two solutions with period π of the form

$$(7.42a) \qquad y_1 = \sum_{0}^{N} A_{2n} \cos 2nx + \sum_{N+1}^{\infty} A_{2n} \cos 2nx$$

and

$$(7.42b) \qquad y_2 = \sum_{1}^{N} B_{2n} \sin 2nx + \sum_{N+1}^{\infty} A_{2n} \sin 2nx$$

or two solutions with period 2π of the form

$$(7.43a) \quad y_1 = \sum_{0}^{N} A_{2n+1} \cos (2n+1)x + \sum_{N+1}^{\infty} A_{2n+1} \cos (2n+1)x$$

and

$$(7.43b) \quad y_2 = \sum_{0}^{N} B_{2n+1} \sin (2n+1)x + \sum_{N+1}^{\infty} A_{2n+1} \sin (2n+1)x$$

cannot exist simultaneously.

We assume, of course, that $C_{2m} \neq 0$.

Proof. Assume that equations (7.42) coexist for $N = 0$. Then by direct substitution in the differential equation we obtain the following pair of identities

$$(7.44a) \quad \sum_{0}^{\infty} [(2C_0 - 8n^2)A_{2n} \cos 2nx$$
$$+ C_2A_{2n} \cos (2n+2)x + \cdots + C_{2m}A_{2n} \cos (2n+2m)x$$
$$+ C_2A_{2n} \cos (2n-2)x + \cdots + C_{2m}A_{2n} \cos (2n-2m)x] = 0,$$

$$(7.44b) \quad \sum_{1}^{\infty} [(2C_0 - 8n^2)A_{2n} \sin 2nx$$
$$+ C_2A_{2n} \sin (2n+2)x + \cdots + C_{2m}A_{2n} \sin (2n+2m)x$$
$$+ C_2A_{2n} \sin (2n-2)x + \cdots + C_{2m}A_{2n} \sin (2n-2m)x] = 0.$$

The recurrence relations obtained from either (7.44a) or (7.44b) will be identical for $n > m$. Hence

$$(7.45) \quad C_{2m}A_{2n+2m} + \cdots + C_2A_{2n+2} + (2C_0 - 8n^2)A_{2n}$$
$$+ C_2A_{2n-2} + \cdots + C_{2m}A_{2n-2m} = 0; \qquad (n > m).$$

Now consider the recurrence relations for $n = m$,

$$\text{(7.46a)} \quad C_{2m}A_{4m} + \cdots + C_2A_{2m+2} + (2C_0 - 8m^2)A_{2m} + C_2A_{2m-2} + \cdots + C_{2m-2}A_2 + C_{2m}A_0 = 0,$$

$$\text{(7.46b)} \quad C_{2m}A_{4m} + \cdots + C_2A_{2m+2} + (2C_0 - 8m^2)A_{2m} + C_2A_{2m-2} + \cdots + C_{2m-2}A_2 = 0.$$

It is clear that $A_0 = 0$.

Comparing the relationships for $n = m - 1$ yields $A_2 = 0$, since $\sin(-2x) = -\sin 2x$ and $\cos(-2x) = \cos 2x$. Thus

$$\text{(7.47a)} \quad C_{2m}A_{4m-2} + \cdots + C_2A_{2m} + [2C_0 - 8(m-1)^2]A_{2m-2} + C_2A_{2m-4} + \cdots + C_{2m-4}A_2 + C_{2m-2}A_0 + C_{2m}A_2 = 0,$$

$$\text{(7.47b)} \quad C_{2m}A_{4m-2} + \cdots + C_2A_{2m} + [2C_0 - 8(m-1)^2]A_{2m-2} + C_2A_{2m-4} + \cdots + C_{2m-4}A_2 - C_{2m}A_2 = 0.$$

But $A_0 = 0$, and therefore, subtracting (7.47b) from (7.47a) gives $2C_{2m}A_2 = 0$.

Continuing this process we get, upon reaching the recurrence relationship for $n = 0$,

$$\text{(7.48)} \qquad 0 = A_0 = A_2 = A_4 = \cdots = A_{2m}.$$

Then with (7.48) and the recurrence relationship for $n = 1$, $A_{2m+2} = 0$ and so on, until the recurrence relationship for $n = m$, $A_{4m} = 0$ is reached. Therefore,

$$\text{(7.49)} \qquad 0 = A_0 = A_2 = \cdots = A_{2m} = \cdots A_{4m}.$$

Now consider equation (7.45) for $n = m + 1$

$$\text{(7.50)} \quad C_{2m}A_{4m+2} + C_{2m-2}A_{4m} + \cdots + C_{2m-2}A_4 + C_{2m}A_2 = 0,$$

and applying equation (7.49)

$$C_{2m}A_{4m+2} = 0.$$

Continuing with $n = m + 2, m + 3, \ldots$, it is clear that all the A's must vanish. Hence the theorem is true for $N = 0$.

If $N \neq 0$, compare the recurrence relationships for $n = N + m$. Since by hypothesis $A_{2N} \neq B_{2N}$ but all $A_{2n} = B_{2n}$ for $n > N + 1$, we obtain

$$\text{(7.51a)} \qquad C_{2m}A_{2N+4m} + \cdots + C_{2m-2}A_{2N+2} + C_{2m}A_{2N} = 0$$

$$\text{(7.51b)} \qquad C_{2m}A_{2N+4m} + \cdots + C_{2m-2}A_{2N+2} + C_{2m}B_{2N} = 0.$$

Subtracting (7.51b) from (7.51a)

$$C_{2m}(A_{2N} - B_{2N}) = 0 \quad \text{or} \quad A_{2N} = B_{2N} \tag{7.52}$$

contrary to hypothesis, and the theorem holds for solutions with period π.

If solutions (7.43) coexist, then for $N = 0$ and $n \geq m$, the recurrence relations are the same for (7.43a) and (7.43b), or

$$C_{2m}A_{2n+2m+1} + \cdots + 2[C_0 - (2n+1)^2]C_{2n+1} + \cdots + C_{2m}A_{2n-2m+1} = 0. \tag{7.53}$$

Compare recurrence relationships for $n = m - 1$ and we find, by the same reasoning as before, that

$$C_{2m}A_1 = 0.$$

By continuing with the relationships for $n = m - 2, m - 3, \ldots$ we obtain

$$0 = A_1 = \cdots = A_{2m-1}. \tag{7.54}$$

Now, from the recurrence relations for $n = 1$, we get $A_{2m+1} = 0$. Examination of the recurrence relations as n increases makes it clear that all the A's must vanish.

For $N \neq 0$, by comparing the recurrence relationships for $n = N + m$, we find the contradiction

$$C_{2m}(A_{2N+1} - B_{2N+1}) = 0, \tag{7.55}$$

which completes the proof of the theorem.

The impossibility of the coexistence of finite trigonometric polynomial solutions, which is not included in Theorem 7.10 could be demonstrated in the same manner. However, it is possible to prove the stronger result:

Theorem 7.11. *Finite symmetric Hill equations cannot have finite trigonometric polynomial solutions.*

Proof. Assume a solution of equation (7.41) exists in the form

$$y = \sum_{0}^{N} A_{2n} \cos 2nx \quad \text{or} \quad y = \sum_{1}^{N} A_{2n} \sin 2nx \tag{7.56}$$

where A_{2N} is the last nonzero coefficient in either of the series (7.56).

For $n > m$ the recurrence relations (7.45) hold. Consider the recurrence relation for $n = N + m$, since all $A_{2n} = 0$ for $n \geq N + 1$

$$C_{2m}A_{2N} = 0 \quad \text{or} \quad A_{2N} = 0. \tag{7.57}$$

This contradicts the hypothesis that A_{2N} is the last nonzero coefficient and proves the theorem for solutions with period π.

Assume that a solution of equation (7.41) exists in the form

$$y = \sum_0^N A_{2N+1} \cos (2N + 1)x \quad \text{or} \quad y = \sum_0^N A_{2N+1} \sin (2N + 1)x, \tag{7.58}$$

where A_{2N+1} is the last nonzero coefficient of either of the series (3). Consider the recurrence relation (7.53) for $n = N + m$. Since all $A_{2n+1} = 0$ for $n > N + 1$

$$C_{2m}A_{2N+1} = 0 \quad \text{or} \quad A_{2N+1} = 0, \tag{7.59}$$

contradicting the hypothesis that A_{2N+1} is the last nonzero coefficient. This proves the theorem.

Further results on finite Hill equations have been found by Winkler (1958).

7.6. *Extreme cases of coexistence*

If all but a finite number of the intervals of instability of Hill's equation

$$y'' + [\lambda + Q(x)]y = 0, \quad Q(x + \pi) = Q(x), \quad \int_0^\pi Q\, dx = 0 \tag{7.60}$$

[with real valued $Q(x)$] vanish, strong statements can be made about $Q(x)$. The first one of these was found by Borg (1946) as a consequence of his theorem as stated in Section 2.6. We shall give here a very simple proof due to Ungar (1961). After that, we shall describe two results obtained by Hochstadt (1965).

There arises a certain difficulty in formulating results which refer to properties of $Q(x)$ deriveable from properties of the discriminant $\Delta(\lambda)$ of (7.60). Unless we confine ourselves to continuous functions $Q(x)$, there obviously exists an infinitude of equations (7.60) with the same $\Delta(\lambda)$. For, if we add to $Q(x)$ a function which vanishes almost everywhere, we obtain the same $\Delta(\lambda)$ as before since $\Delta(\lambda)$ depends only on the solutions of (7.60) which are, by definition, twice differentiable functions of x and analytic in λ. Therefore, the least we will have to assume about $Q(x)$ is that Q is differentiable and is the integral of its derivative. Assuming then throughout this section that $Q(x)$ belongs to the class of functions thus described, we have

Theorem 7.12. *If Hill's equation* (7.60) *is such that all intervals of instability disappear, with the exception of the zeroth interval* $(-\infty, \lambda_0)$, *then* $Q(x) \equiv 0$.

Proof. According to Theorem 2.1 and Corollary 2.1, $\Delta^2(\lambda) - 4$ has double roots $\lambda_{2n-1} = \lambda_{2n}$ and $\lambda'_{2n-1} = \lambda'_{2n}$ for $n = 1, 2, 3, \ldots$, and a simple root $\lambda = \lambda_0$. Since it has no other zeros, the function

$$F(\lambda) = \tfrac{1}{2}[\Delta(\lambda) + (\Delta^2 - 4)^{1/2}]$$

has exactly one branch point of order 2 at $\lambda = \lambda_0$ and is analytic everywhere else. Therefore,

$$G(z) = F(z^2 + \lambda_0)$$

is an entire function of z [more precisely, one of two entire functions, according to the sign of the square root in the definition of $F(\lambda)$]. $G(z)$ cannot vanish anywhere since

$$[\Delta(\lambda) + (\Delta^2 - 4)^{1/2}][\Delta(\lambda) - (\Delta^2 - 4)^{1/2}] = 4.$$

Therefore $\log G(z)$ is again an entire function of z. According to Theorem 2.2 [equation (2.35)], we have for its real part:

$$\operatorname{Re} \log G(z) \leq \pi|z| + K,$$

where K is a constant. Hence we have

$$\log G(z) = az + b,$$

with constant a and b and

$$\Delta(\lambda) = 2A \cosh az = 2A \cosh a\,(\lambda - \lambda_0)^{1/2},$$

where again A is a constant. We apply the first statement of Theorem 2.4 (Section 2.2) and obtain

$$a = \pm i\pi, \quad A = 1, \quad \lambda_0 = 0.$$

Now we conclude from Theorem 4.4 that $Q(x) \equiv 0$.

Hochstadt (1965) developed a new method for proving Theorem 7.12 which also yields more general results. We state without proof his two theorems which are no longer contained in any older results. We have

Theorem 7.13. *If Hill's equation* (7.60) *is such that all but two of its intervals of instability disappear, then $Q(x)$ satisfies a differential equation*

$$Q'' = 3Q^2 + AQ + B$$

or

$$\{Q'\}^2 = 2Q^3 + AQ^2 + 2BQ + C,$$

where A, B, and C are constants. Therefore, Q is a Weierstrass elliptic function, and by a linear transformation of both dependent and independent variable (7.60), *can be transformed into Lamé's equation* (7.31).

Since Lamé's equation has been discussed in all details (see Section 7.3), Theorem 7.13 settles completely the question for which $Q(x)$ only two intervals of instability can remain. Finally, Hochstadt proved

Theorem 7.14. *If $Q(x)$ is such that* (7.60) *has only a finite number of intervals of instability, then $Q(x)$ is infinitely differentiable.*

VIII

Examples

In this chapter, we shall discuss briefly a few examples of equations of Hill's type which are particularly easily approachable and which may be useful for the purpose of a first orientation. The first three cases are distinguished by the fact that they admit an explicit integration of Hill's equation in terms of well-known functions. The fourth example (the equation for frequency modulation) offers the simplest illustration available for the results on coexistence in Chapter VII.

8.1. Impulse functions

In connection with a problem in quantum theory (electrons in a one-dimensional conductor), Kronig and Penney (1931) studied the special equation of Hill's type:

$$y'' + [\lambda + Q(x)]y = 0, \tag{8.1}$$

where

$$\begin{aligned} Q(x) &= -v_0 \quad \text{for} \quad -b < x < 0, \\ Q(x) &= 0 \quad \text{for} \quad 0 < x < a, \\ Q(x + c) &= Q(x) \quad \text{where} \quad c = a + b. \end{aligned}$$

The problem is to find intervals of stability, that is, those values of λ for which (8.1) has a nontrivial solution y with the property

$$y(x + c) = e^{ikc}y(x)$$

where k is real. Such a solution will exist if and only if the equation

$$\cos(a\sqrt{\lambda})\cos(b\sqrt{\lambda - v_0}) + \frac{v_0 - 2\lambda}{\sqrt{\lambda}\sqrt{\lambda - v_0}}\sin(a\sqrt{\lambda})\sin(b\sqrt{\lambda - v_0}) = \cos kc \tag{8.2}$$

can be satisfied for a real value of k. Of course, this means that the left-hand side of (8.2) must lie between -1 and $+1$.

Equation (8.2) is still somewhat hard to discuss numerically. A further simplification can be obtained if we replace Q (which so far has the shape of a "well") by a Dirac delta function. This can be done by letting

$$b \to 0, \quad v_0 \to \infty, \quad bv_0 \to P/a,$$

where P is a positive constant. Then (8.2) reduces to

$$\cos a\sqrt{\lambda} + P/(a\sqrt{\lambda}) \sin a\sqrt{\lambda} = \cos kc, \tag{8.3}$$

and the condition that the left-hand side of (8.3) should have values between -1 and $+1$ can be discussed readily by drawing the curve

$$u = \cos v + (P/v) \sin v$$

in the (u, v) plane. It follows that the intervals of instability [where (8.3) cannot hold for a real value of k] will tend towards zero like $\lambda^{-1/2}$ as $\lambda \to +\infty$.

We followed here the exposition given by Sommerfeld and Bethe (1933) where many more details may be found. See also Strutt (1932).

8.2. *Piecewise constant functions*

Meissner (1918) investigated the equation

$$y'' + \omega^2 q(x) y = 0 \tag{8.4}$$

where $q(x)$ assumes a finite number of different values in the interval $0 \leq x \leq 2\pi$ and is periodic with period 2π. Although (8.4) cannot be transformed into an equation of type (8.1) since the Liouville transformation is not applicable to discontinuous functions, equation (8.4) is relevant to the theory of Hill's equation and we shall now discuss it briefly.

Assume that the interval $0 \leq x \leq 2\pi$ has been divided into n parts of length τ_i, $i = 1, \ldots, n$, $\tau_1 + \cdots + \tau_n = 2\pi$, and assume that

$$q(x) = v_i^2/(4\pi^2)$$

in the ith interval. We shall use the notations

$$C_i = \cos(\omega v_i \tau_i/2\pi), \quad S_i = \sin(\omega v_i \tau_i/2\pi),$$

$$V_{i,k} = \frac{1}{2}\left(\frac{v_i}{v_k} + \frac{v_k}{v_i}\right), \qquad (i, k = 1, 2, \ldots, n),$$

$$J_3 = C_1C_2C_3 - v_{12}S_1S_2C_3 - v_{13}S_1S_3C_2 - v_{23}S_2S_3C_1,$$

$$J_2 = C_1C_2 - v_{12}S_1S_2.$$

Then the necessary and sufficient condition for (8.4) to have a periodic solution of period 2π is, for $n = 3$, $J_3 = 1$ and for $n = 2$, $J_2 = 1$. Similarly, (8.4) will have a periodic solution of period 4π if and only if $J_3 = -1$ (for $n = 3$) or $J_2 = -1$ (for $n = 2$). If, in particular,

$$\tau_1 = \tau_2 = \pi,$$

and if we put

$$x_1 = \omega v_1/2, \quad x_2 = \omega v_2/2,$$

we find

$$J_2 = J_2(x_1, x_2) = \cos x_1 \cos x_2 - \frac{1}{2}\left(\frac{x_1}{x_2} - \frac{x_2}{x_1}\right) \sin x_1 \sin x_2. \tag{8.5}$$

Meissner draws the curves $J_2 = \pm 1$ in the x_1, x_2 plane. They consist of infinitely many separate branches bounding the regions of instability for (8.4) in the case $n = 2$.

Hochstadt (1960) analysed more thoroughly an equation which is equivalent to (8.4). Considering

$$y'' + \omega^2 Q(x) y = 0 \tag{8.6}$$

with

$$Q(-x) = Q(x), \quad Q(x + 2L) = Q(x),$$

$$Q(x) = 1 \quad \text{for} \quad |x| \leq 1, \qquad Q(x) = a \quad \text{for} \quad 1 < |x| \leq L,$$

he shows that (8.6) will have a periodic solution of period $2L$ for ω_m, $m = 0, 1, 2, \ldots$, where, for large $m = 2n - 1$ or $m = 2n$,

$$\omega_{2n} = \left[\frac{2n}{1 + a(L - 1)}\right]\frac{\pi}{2} + \theta\,\frac{\pi}{2},$$

$$\omega_{2n-1} = \left[\frac{2n}{1 + a(L - 1)}\right]\frac{\pi}{2} + \theta^*\,\frac{\pi}{2}.$$

Here $[u]$ denotes the largest integer not exceeding u, and θ and θ^* are undetermined quantities satisfying the inequalities

$$0 \le \theta,\ \theta^* < 1$$

Similarly, (8.6) will have solutions of period $4L$ if $\omega = \omega'_m$, where, for large $m = 2n$ or $m = 2n - 1$,

$$\omega'_{2n} = \left[\frac{2n - 1}{1 + a(L - 1)}\right]\frac{\pi}{2} + \theta\,\frac{\pi}{2},$$

$$\omega'_{2n-1} = \left[\frac{2n - 1}{1 + a(L - 1)}\right]\frac{\pi}{2} + \theta^*\,\frac{\pi}{2}.$$

It can be shown that the lengths of the intervals of instability, that is, the differences

$$|\omega^2_{2n} - \omega^2_{2n-1}|, \quad |\omega'^2_{2n} - \omega'^2_{2n-1}|$$

will, in general, tend to infinity as $n \to \infty$. Also, coexistence of periodic solutions of period $2L$ or $4L$ can be shown to take place if and only if $a(L - 1)$ is a rational number.

8.3. *Piecewise linear reciprocal function*

Schwerin (1931) also investigated the case where $Q(x)$ is piecewise constant and, in addition, the case where $1/Q(x)$ is a piecewise linear function, and where $\lambda = 0$. In this case, (8.1) can be integrated explicitly in terms of Bessel functions. Using the notation of Schwerin's paper, we shall write the differential equation in the form

$$\frac{d^2\eta}{dt^2} + \theta^2\,\frac{\eta}{\varepsilon(t)} = 0 \tag{8.7}$$

where $\varepsilon(t)$ is either a triangular function:

$$\varepsilon(t) = \varepsilon_0 + 2(\varepsilon_1 - \varepsilon_0)t \quad \text{for} \quad 0 \le t \le \tfrac{1}{2},$$

$$\varepsilon(t) = \varepsilon_1 - 2(\varepsilon_1 - \varepsilon_0)(t - \tfrac{1}{2}) \quad \text{for} \quad \tfrac{1}{2} \le t \le 1,$$

$$\varepsilon(t + 1) = \varepsilon(t), \quad 0 < \varepsilon_0 < \varepsilon_1, \quad \varepsilon_0 + \varepsilon_1 = 2.$$

or a wedge function: (with a jump at $t = 0, \pm 1, \pm 2, \ldots$)

$$\varepsilon(t) = \varepsilon_0 + (\varepsilon_1 - \varepsilon_0)t \quad \text{for} \quad 0 < t < 1.$$

$$\varepsilon(t + 1) = \varepsilon(t), \quad 0 < \varepsilon_0 < \varepsilon_1, \quad \varepsilon_0 + \varepsilon_1 = 2.$$

In the case of the triangular function, the values of θ for which (8.7) has a periodic solution of period 1 are the roots of a transcendental function which is defined as follows:

Let J_m and Y_m $(m = 0, 1)$ be, respectively, the Bessel and the Neumann function of order m. Then θ must satisfy one of the equations

$$J_1\left(\frac{\theta\sqrt{\varepsilon_0}}{\varepsilon_1 - \varepsilon_0}\right)Y_1\left(\frac{\theta\sqrt{\varepsilon_1}}{\varepsilon_1 - \varepsilon_0}\right) = J_1\left(\frac{\theta\sqrt{\varepsilon_1}}{\varepsilon_1 - \varepsilon_0}\right)Y_1\left(\frac{\theta\sqrt{\varepsilon_0}}{\varepsilon_1 - \varepsilon_0}\right)$$

or

$$J_0\left(\frac{\theta\sqrt{\varepsilon_0}}{\varepsilon_1 - \varepsilon_0}\right)Y_0\left(\frac{\theta\sqrt{\varepsilon_1}}{\varepsilon_1 - \varepsilon_0}\right) = J_0\left(\frac{\theta\sqrt{\varepsilon_1}}{\varepsilon_1 - \varepsilon_0}\right)Y_0\left(\frac{\theta\sqrt{\varepsilon_0}}{\varepsilon_1 - \varepsilon_0}\right)$$

if (8.7) has a solution of period 1. Similarly, (8.7) will have a solution of period 2 if and only if one of the following equations is satisfied:

$$J_0\left(\frac{\theta\sqrt{\varepsilon_0}}{\varepsilon_1 - \varepsilon_0}\right)Y_1\left(\frac{\theta\sqrt{\varepsilon_1}}{\varepsilon_1 - \varepsilon_0}\right) = J_1\left(\frac{\theta\sqrt{\varepsilon_1}}{\varepsilon_1 - \varepsilon_0}\right)Y_0\left(\frac{\theta\sqrt{\varepsilon_0}}{\varepsilon_1 - \varepsilon_0}\right)$$

or

$$J_1\left(\frac{\theta\sqrt{\varepsilon_0}}{\varepsilon_1 - \varepsilon_0}\right)Y_0\left(\frac{\theta\sqrt{\varepsilon_1}}{\varepsilon_1 - \varepsilon_0}\right) = J_0\left(\frac{\theta\sqrt{\varepsilon_1}}{\varepsilon_1 - \varepsilon_0}\right)Y_1\left(\frac{\theta\sqrt{\varepsilon_0}}{\varepsilon_1 - \varepsilon_0}\right)$$

Similar equations can be obtained if $\varepsilon(t)$ is a wedge-shaped function. Schwerin (1931) draws stability charts showing the regions of stability and instability in a plane with the Cartesian coordinates θ and ε_0. Also, numerical tables and a large variety of comments (in part empirical) on the regions of stability may be found in this paper.

8.4. *The frequency modulation equation*

The equation

$$(8.8) \qquad (1 + a\cos 2x)y'' + \lambda y = 0$$

has been studied by Cambi (1948, 1949). Because of its applications, it may be called the equation of frequency modulation. We shall assume that $|a| < 1$ and, of course, that a and λ are real. Equation (8.8) is not in the form of (8.1), but it can be transformed into it by using the Liouville transformation of Section 3.1. Therefore, we may talk about intervals of stability and instability on the λ axis. According to Corollary 4.2, the zeroth interval of instability contains all values $\lambda < 0$. Obviously, $\lambda_0 = 0$, since for $\lambda = 0$, (8.8) has the periodic solution $y = 1$.

We shall now prove that, except for the zeroth one, all even intervals of instability disappear for (8.8) but no odd interval of instability disappears if $a \neq 0$. If we look at Theorem 7.1, we see at once that $Q^*(\mu) = a(2\mu - 1)^2$ will not vanish for any integral value of μ, unless $a = 0$, in which case (8.8) is trivial. This proves our statement about the odd-numbered intervals of instability. With respect to the even ones, we find that $Q(\mu)$, as defined in Theorem 7.1, is simply $a\mu^2$ and therefore has the integral root zero and no other one (if $a \neq 0$). Now Theorem 7.3 shows that (8.8) must have two linearly independent solutions of period π whenever it has one such solution unless this solution is a constant. However, a constant solution of (8.8) can occur only for $\lambda = 0$, a case already considered.

List of Symbols and Notations

Q	coefficient in Hill's equation (except for Chapter VII): page 3
y_1, y_2	normalized solutions: page 3
g_n	Fourier coefficients of Q: page 24
$\operatorname{sgn} n$	sign of n: page 35
$\delta_{n,m}$	Kronecker symbol: page 28
$\lambda_0, \lambda_1, \ldots$	characteristic values of the first kind: page 11
$\lambda_1', \lambda_2', \ldots$	characteristic values of the second kind: page 11
$\Delta(\lambda)$	discriminant: page 15
Δ_n	homogeneous components of Δ: page 24

List of Theorems, Lemmas, and Corollaries

References

Atkinson, F. V. (1957), Estimation of an Eigenvalue Occurring in a Stability Problem, *Math. Z.*, **68**, 82–99.

Birkhoff, G. D. (1909), Existence and Oscillation Problems for a Certain Boundary Value Problem, *Trans. Am. Math. Soc.*, **10**, 259–270.

Blumenson, L. E. (1963), On the Eigenvalues of Hill's Equation, *Comm. Pure Appl. Math.*, **16**, 261–266.

Borg, G. (1944), Über die Stabilität gewisser Klassen von linearen Differentialgleichungen, *Ark. Mat. Astr. Fys.*, **31A**, No. 1, pp. 1–3.

Borg, G. (1946), Eine Umkehrung der Sturm-Liouvilleschen Eigenwertaufgabe. Bestimmung der Differentialgleichung durch die Eigenwerte, *Acta Math.*, **78**, 1–96.

Borg, G. (1949), On a Liapounoff Criterion of Stability, *Amer. J. Math.*, **71**, 67–70.

Brillouin, L. (1946), Wave Propagation in Periodic Structures, in *Electric Filters and Crystal Lattices*, McGraw-Hill, New York.

Brillouin, L. (1948), A Practical Method for Solving Hill's Equation, *Quart. Applied Math.*, **6**, 167–178.

Brillouin, L. (1950), The B.W.K. Approximation and Hill's Equation, II, *Quart. Appl. Math.*, **7**, 363–380.

Cambi, E. (1948), Trigonometric Components of a Frequency Modulated Wave, *Proc. I.R.E.*, **36**, 44–49.

Cambi, E. (1949), The Simplest Form of Second Order Linear Differential Equations with Periodic Coefficients Having Finite Singularities, *Proc. Roy. Soc. Edinburgh*, A **63**, 27–51 (1949).

Cesari, L. (1959), Asymptotic Behavior and Stability Problems in Ordinary Differential Equations, *Ergebnisse der Mathematik* (*New series*), No. 16, 271 pp. Springer, Berlin–Goettingen–Heidelberg.

Coddington, E. A., and N. Levinson (1955), *Theory of Ordinary Differential Equations*, McGraw-Hill, New York.

Courant, R., and D. Hilbert (1953), *Methods of Mathematical Physics*, Vol. 1, Chap. 5, Interscience Publishers, New York.

Erdélyi, A. (1934), Ueber die freien Schwingungen in Kondensatorkreisen von veraenderlicher Kapazitaet, *Annalen der Physik*, (5), **19**, 585–622.

Erdélyi, A. (1935), Ueber die rechnerisch Ermittlung der Schwingungsvorgaenge in Kreisen mit periodisch schwankenden Parametern; *Archiv fuer Elektrotechnik*, **29**, 473–489.

Erdélyi, A. (1941), On Lamé functions, *Philos. Mag.*, (7), **31**, 123–130; **32**, 348–350.

Erdélyi, A., et al. (1955), *Higher Transcendental Functions*, Vol. 3, McGraw-Hill, New York.

Gambill, R. A. (1954), Stability Criteria for Linear Differential Systems with Periodic Coefficients, *Riv. Mat. Univ. Parma*, **5**, 169–181.

Gambill, R. A. (1955), Criteria for Parametric Instability for Linear Differential Systems with Periodic Coefficients, *Riv. Mat. Univ. Parma*, **6**, 37–43.

Gel′fand, I. M., and B. M. Levitan (1951), On the Determination of a Differential Equation from its Spectral Function, *Izv. Akad. Nauk SSSR, Ser. Math.*, **15**, 309, §11, (Russian), Engl. trans. in *Amer. Math. Soc. Translations*, Series 2, Vol. 1, Providence, R.I., 1955.

Gol′din, A. M. (1951), On a Criterion of Lyapunov, *Akad. Nauk SSR, Prikl. Mat. Meh.*, **15**, 379–384.

Golomb, M. (1958), Expansion and Boundedness Theorems for Solutions of Linear Differential Systems with Periodic or Almost Periodic, Coefficients, *Arch. Rat. Mech. and Anal.*, **2**, 284–308.

Haacke, W. (1952), Ueber die Stabilitaet eines Systems von gewoehnlichen Differentialgleichungen zweiter Ordnung mit periodischen Koëffizienten, die von Parametern abhaengen, *Math. Z.*, **56**, 65–79; **57**, 34–45.

Hamel, G. (1913), Über die Differentialgleichungen zweiter Ordnung mit periodischen Koëffizienten, *Math. Ann.*, **73**, 371–412.

Hardy, G. H., J. E. Littlewood, and G. Polya (1934), *Inequalities*, Cambridge.

Hartman, P., and A. Wintner (1949), On the Location of Spectra of Wave Equations, *Amer. J. Math.*, **71**, 214–217.

Haupt, O. (1914), Über eine Methode zum Beweis von Oszillationstheoremen, *Math. Ann.*, **76**, 67–104.

Haupt, O. (1919), Über lineare homogene Differentialgleichungen 2. Ordnung mit periodischen Koëffizienten, *Math. Ann.*, **79**, 278–285.

Hill, G. W. (1886), On the Part of the Motion of the Lunar Perigee which is a Function of the Mean Motions of the Sun and Moon, *Acta Math.*, **8**, 1–36; reprinted, with some additions, from a paper published at Cambridge, U.S.A., 1877.

Hochstadt, H. (1960), A Special Hill's Equation with Discontinuous Coefficients, N.Y.U. Inst. Math. Sci., Div. of EM Res., Report BR-32.

Hochstadt, H. (1961), Asymptotic Estimates for the Sturm-Liouville Spectrum, *Comm. Pure Appl. Math.*, **14**, 749–764.

Hochstadt, H. (1962), A Class of Stability Criteria for Hill's Equation, *Quart. Appl. Math.*, **20**, 92–93.

Hochstadt, H. (1963a), Function Theoretic Properties of the Discriminant of Hill's Equation, *Math. Z.*, **82**, 237–242.

Hochstadt, H. (1963b), Estimates on the Stability Intervals for Hill's Equation, *Proc. American Math. Soc.*, **14**, 930–932.

Hochstadt, H. (1963c), On the Asymptotic Spectrum of Hill's Equation, *Archiv d. Mathematik*, **14**, 34–38.

Hochstadt, H. (1965), On the Characterization of a Hill's Equation via its Spectrum, *Archive for Rational Mechanics and Analysis*, **19**, 353–362.

Humbert, P. (1926a), Some Hyperspace Harmonic Analysis Problems Introducing Extensions of Mathieu's Equations; *Proc. Roy. Soc. Edin.*, **46**, 206–209.

Humbert, P. (1926b), Fonctions de Lamé et fonctions de Mathieu, *Mémorial des Sciences Mathématiques, Fasc.*, 10.

Ince, E. L. (1922), A Proof of the Impossibility of the Coexistence of Two Mathieu Functions, *Proc. Camb. Philos. Soc.*, **21**, 117–120.

Ince, E. L. (1923), A Linear Differential Equation with Periodic Coefficients, *Proc. Lond. Math. Soc.* (2), **23**, 56–74.

Ince, E. L. (1925), The Modes of Vibration of a Stretched Membrane with a Particular Law of Density, *Proc. Roy. Soc. Edin.*, **45**, 102–116.

Ince, E. L. (1926), Periodic Solutions of a Linear Differential Equation of the Second Order, *Proc. Camb. Philos. Soc.*, **23**, 44–46; The Real Zeros of Solutions of a Linear Differential Equation with Periodic Coefficients, *Proc. Lond. Math. Soc.* (2), **25**, 53–58.

Ince, E. L. (1940), The Periodic Lamé Functions, *Proc. Roy. Soc. Edin.*, **60**, 47–63; Further Investigations into the Periodic Lamé Functions, *Proc. Roy. Soc. Edin.*, **60**, 83–99.

Ince, E. L. (1944), *Ordinary Differential Equations*, Dover, New York.

Jagerman, D. L. (1962), *The Discriminant of Hill's Equation*, New York University, Courant Institute of Math. Sciences, Div. of Electromagnetic Research, Report BR-39.

Klotter, K. and G. Kotowski (1943), Ueber die Stabilitaet von Loesungen Hillscher Differentialgleichungen mit drei unabhaengigen Parametern, *Z. angew. Math. Mech.*, **23**, 149–155.

Krein, M. G. (1951), On Certain Problems on the Maximum and Minimum of Characteristic Values and on the Lyapunov Zones of Stability, *Prikl. Mat. Meh. 15*, 323–348, (Russian), Engl. trans. in *Amer. Math. Soc. Translations*, Series 2, Vol. 1, Providence, R.I., 1955.

Kronig, R. de L. and W. G. Penney (1931), Quantum Mechanics in Crystal Lattices, *Proc. Royal Soc. London*, **130**, 499–513.

Levin, B. Ja. (1964), *Distribution of Zeros of Entire Functions*, American Mathematical Society, Providence, R.I., Translations of Mathematical Monographs Vol. 5, Sect. 3, Appendix 1, Payley-Wiener Theorem, Sect. 2, Appendix 4, Inverse Sturm-Liouville Problem.

Levy, Dorothy M. and J. B. Keller (1963), Instability Intervals of Hill's equation, *Comm. Pure Appl. Math.*, **16**, 469–476.

Liapounoff, A. (1902), Sur une série dans la théorie des equations différentielles linéaires du second ordre à coefficients périodiques, *Zap. Imp. Akad. Nauk. Fiz.-Mat. Otd.*, **13**, No. 2.

Liapounoff, A. (1907), Problème général de la stabilité du movement, *Ann. Fac. Sci. Univ. Toulouse* (2), **9**, 203–474. Repr. Princeton Univ. Press.

Magnus, W. (1955), Infinite Determinants Associated with Hill's Equation, *Pacific Journal of Math.*, **5**, Suppl. 2, 941–951.

Magnus, W. (1959), *The Discriminant of Hill's Equation*, N.Y.U. Inst. Math. Sci., Div. EM Res., Report BR-28.

McLachlen, N. W. (1947), *Theory and application of Mathieu Functions*, Clarendon Press, Oxford.

Meissner, E. (1918), Ueber Schuettelschwingungen in Systemen mit periodisch veraenderlicher Elastizitaet, *Schweizer Bauzeitung*, **72**, No. 10, 95–98.

Meixner, J. and F. W. Schaefke (1954), *Mathieusche Funktionen und Sphaeroid Funktionen*, Grundlehren der Mathematischen Wissenschaften, **71**, Springer.

Moulton, F. R. (1930), *Differential Equations*, Macmillan, New York.

Nevanlinna, R. (1936), *Eindeutige analytische Funktionen*, Springer.

Putnam, C. R. (1954), On the Gaps in the Spectrum of the Hill Equation, *Quart. Appl. Math.*, **11**, 496–498.

Schwerdtfeger, H. (1946), The Eigenvalue Problem of Hill's Equation, *Journal and Proceedings of the Royal Society of New South Wales*, **79**, 176–189.

Schaefke, F. W. (1960/61) Reihenentwicklungen analytischer Funktionen nach Biorthogonalsystemen spezieller Funktionen I, II, *Math. Zeitschrift*, **74**, 436–470; **75**, 151–191.

Schaefke, F. W., see Meixner.

Schwerin, E. (1931), Ein allgemeines Integrationsverfahren fuer quasiharmonische Schwingungsvorgaenge, *Zeitsch. fuer Techn. Physik*, **12**, 104–111.

Sommerfeld, A. and H. Bethe (1933), Elektronen Theorie der Metalle, *Handbuch der Physik*, 2nd Ed., Vol. 24[II], 379–385.

Starzinskiĭ, V. M. (1954), Survey of Works on Conditions of Stability of the Trivial Solution of a System of Linear Differential Equations with Periodic Coefficients, *Prikl. Mat. Meh. 18*, 469–510, (Russian), Engl. trans. in *Amer. Math. Soc. Translations*, Series 2, Vol. 1, Providence, R.I., 1955.

Strutt, M. J. O. (1932), Lamésche, Mathieusche und verwandte Funktionen, *Ergebnisse der Math.*, **1**, 3, Springer.

Strutt, M. J. O. (1943), Grenzen voor de Eigenwaarden bij Problemen van Hill I, II, Eigenfuncties bij Problem van Hill I, II, Eigenwaarde-Krommen bij Problemen van Hill I, II, *Nederland. Akad. van Wetenschappen*, **52**, Nos. 8, 9, 2, 3, 4, 5.

Strutt, M. J. O. (1944), Reelle Eigenwerte verallgemeinerter Hillscher Eigenwertaufgaben 2. Ordnung, *Math. Z.*, **49**, 593–643.

Strutt, M. J. O. (1948), On Hill's Pr[illegible] with Complex Parameters and a Real Periodic Solution [illegible] *Edin.*, **62**, 278–296.

Titchm[illegible] Hill Equations [illegible] Problems with Periodic Potentials, *Proc.* [illegible], **203**, 501–514.

[illegible] Über gewöhnliche Differentialgleichungen mit Singularitäten und zugehörigen Entwicklungen willkürlicher Funktionen, *Math. Ann.*, [illegible], 220–269.

Whittaker, E. T. (1914), On the General Solution of Mathieu's Equation, *Proc. Edinburgh Math. Soc.*, **32**, 75–80.

Whittaker, E. T. (1915), On a Class of Differential Equations whose Solutions Satisfy Integral Equations, *Proc. Edinburgh Math. Soc.*, **33**, 14–33.

Whittaker, E. J. and G. N. Watson (1927), *A Course of Modern Analysis*, University Press, Cambridge.

Winkler, S. and W. Magnus (1958), *The Coexistence Problem for Hill's Equation*, N.Y.U. Inst. Math. Sc., Div. EM Res., Report BR-26.

[illegible]chin, M. (1946), Sur les conditions pour qu'une solution d'un système linéaire du second ordre possède deux zéros, *C.R. (Doklady) Acad. Sci. U.S.S.R. (N.S.)*, **51**, 573–576.

Meixner, J. and F. W. Schaefke (1954), *Mathieusche Funktionen und Sphaeroid Funktionen*, Grundlehren der Mathematischen Wissenschaften, **71**, Springer.

Moulton, F. R. (1930), *Differential Equations*, Macmillan, New York.

Nevanlinna, R. (1936), *Eindeutige analytische Funktionen*, Springer.

Putnam, C. R. (1954), On the Gaps in the Spectrum of the Hill Equation, *Quart. Appl. Math.*, **11**, 496–498.

Schwerdtfeger, H. (1946), The Eigenvalue Problem of Hill's Equation, *Journal and Proceedings of the Royal Society of New South Wales*, **79**, 176–189.

Schaefke, F. W. (1960/61) Reihenentwicklungen analytischer Funktionen nach Biorthogonalsystemen spezieller Funktionen I, II, *Math. Zeitschrift*, **74**, 436–470; **75**, 151–191.

Schaefke, F. W., see Meixner.

Schwerin, E. (1931), Ein allgemeines Integrationsverfahren fuer quasiharmonische Schwingungsvorgaenge, *Zeitsch. fuer Techn. Physik*, **12**, 104–111.

Sommerfeld, A. and H. Bethe (1933), Elektronen Theorie der Metalle, *Handbuch der Physik*, 2nd Ed., Vol. 24^{II}, 379–385.

Starzinskiĭ, V. M. (1954), Survey of Works on Conditions of Stability of the Trivial Solution of a System of Linear Differential Equations with Periodic Coefficients, *Prikl. Mat. Meh. 18*, 469–510, (Russian), Engl. trans. in *Amer. Math. Soc. Translations*, Series 2, Vol. 1, Providence, R.I., 1955.

Strutt, M. J. O. (1932), Lamésche, Mathieusche und verwandte Funktionen, *Ergebnisse der Math.*, **1**, 3, Springer.

Strutt, M. J. O. (1943), Grenzen voor de Eigenwaarden bij Problemen van Hill I, II, Eigenfuncties bij Problem van Hill I, II, Eigenwaarde-Krommen bij Problemen van Hill I, II, *Nederland. Akad. van Wetenschappen*, **52**, Nos. 8, 9, 2, 3, 4, 5.

Strutt, M. J. O. (1944), Reelle Eigenwerte verallgemeinerter Hillscher Eigenwertaufgaben 2. Ordnung, *Math. Z.*, **49**, 593–643.

Strutt, M. J. O. (1948), On Hill's Problem with Complex Parameters and a Real Periodic Solution, *Proc. Roy. Soc. Edin.*, **62**, 278–296.

Titchmarsh, E. C. (1950), Eigenfunction Problems with Periodic Potentials, *Proc. Royal Soc. London Ser. A.*, **203**, 501–514.

Ungar, P. (1961), Stable Hill Equations, *Comm. Pure Appl. Math.*, **14**, 707–710.

Weyl, H. (1910), Uber gewöhnliche Differentialgleichungen mit Singularitäten und die zugehörigen Entwicklungen willkürlicher Funktionen, *Math. Ann.*, **67**, 220–269.

Whittaker, E. T. (1914), On the General Solution of Mathieu's Equation, *Proc. Edinburgh Math. Soc.*, **32**, 75–80.

Whittaker, E. T. (1915), On a Class of Differential Equations whose Solutions Satisfy Integral Equations, *Proc. Edinburgh Math. Soc.*, **33**, 14–33.

Whittaker, E. J. and G. N. Watson (1927), *A Course of Modern Analysis*, University Press, Cambridge.

Winkler, S. and W. Magnus (1958), *The Coexistence Problem for Hill's Equation*, N.Y.U. Inst. Math. Sc., Div. EM Res., Report BR-26.

Yelchin, M. (1946), Sur les conditions pour qu'une solution d'un système linéaire du second ordre possède deux zéros, *C.R.* (*Doklady*) *Acad. Sci. U.S.S.R.* (*N.S.*), **51**, 573–576.

Additional References (1978)

Flaschka, H (1975), On the Inverse Problem for Hill's Operator, *Arch. Rat. Mech. and Anal.*, **59,** 293–304.

Goldberg, W. (1975), Hill's Equation for a Finite Number of Instability Intervals, *J. Math. Anal. and Appl.*, **51,** 705–723.

Goldberg, W. (1976), Necessary and Sufficient Conditions for Determining a Hill's Equation from Its Spectrum, *J. Math. Anal. and Appl.*, **55,** 549–554.

Hochstadt, H. (1966), On the Determination of a Hill's Equation from Its Spectrum, *II. Arch. Rat. Mech. and Anal.*, **23,** 237–238.

Hochstadt, H. (1968), Hill's Equation with Halfperiodic Coefficients, *Proc. Amer. Math. Soc.*, **19,** 483–486.

Hochstadt, H. (1969), On an Inequality of Lyapunov, *Proc. Amer. Math. Soc.*, **22,** 282–284.

Hochstadt, H. (1975), On the Theory of Hill's Matrices and Related Inverse Spectral Problems, *Linear Algebra & Appl.*, **11,** 41–52.

Hochstadt, H. (1976a), An Inverse Problem for Hill's Equation, *J. Differential Equa.*, **20,** 53–60.

Hochstadt, H. (1976b), An Inverse Problem for Hill's Equation, *SIAM J. J. Appl. Math.*, **31,** 392–396.

Hochstadt, H. (1977), On a Hill's Equation with Double Eigenvalues, *Proc. Amer. Math. Soc.*, **65,** 373–374.

Hochstadt, H. (1978), A Direct and Inverse Problem for a Hill's Equation with Double Eigenvalues. To appear in *J. Math. Anal. Appl.*

Lax, P. (1975), Periodic Solutions of the KdV Equation, *Comm. Pure Appl. Math.*, **28,** 141–188.

Magnus, W. (1976), Monodromy Groups and Hill's Equation, *Comm. Pure Appl. Math.*, **29,** 701–716.

McKean, H. P., and P. van Moerbecke (1975), The Spectrum of Hill's Equation, *Inventiones Math.*, **30,** 217–274.

McKean, H. P., and E. Trubowitz (1976), Hill's Operator and Hyperelliptic Function Theory in the Presence of Infinitely Many Branch Points. *Comm. Pure Appl. Math.*, **29,** 143–226.

Trubowitz, E. (1977), The Inverse Problem for Periodic Potentials, *Comm. Pure Appl. Math.*, **30,** 321–337.

Index

1134